Balarabe Inuwa Dutse
Mustapha Mukhtar

An Empirical Study of Determinants of Food Insecurity in Jigawa
State

Balarabe Inuwa Dutse
Mustapha Mukhtar

An Empirical Study of Determinants of Food Insecurity in Jigawa State

Food Insecurity in Nigeria

LAP LAMBERT Academic Publishing

Impressum/Imprint (nur für Deutschland/only for Germany)
Bibliografische Information der Deutschen Nationalbibliothek: Die Deutsche Nationalbibliothek verzeichnet diese Publikation in der Deutschen Nationalbibliografie; detaillierte bibliografische Daten sind im Internet über http://dnb.d-nb.de abrufbar.
Alle in diesem Buch genannten Marken und Produktnamen unterliegen warenzeichen-, marken- oder patentrechtlichem Schutz bzw. sind Warenzeichen oder eingetragene Warenzeichen der jeweiligen Inhaber. Die Wiedergabe von Marken, Produktnamen, Gebrauchsnamen, Handelsnamen, Warenbezeichnungen u.s.w. in diesem Werk berechtigt auch ohne besondere Kennzeichnung nicht zu der Annahme, dass solche Namen im Sinne der Warenzeichen- und Markenschutzgesetzgebung als frei zu betrachten wären und daher von jedermann benutzt werden dürften.

Coverbild: www.ingimage.com

Verlag: LAP LAMBERT Academic Publishing GmbH & Co. KG
Heinrich-Böcking-Str. 6-8, 66121 Saarbrücken, Deutschland
Telefon +49 681 3720-310, Telefax +49 681 3720-3109
Email: info@lap-publishing.com

Approved by: Bayero University, Kano, Dissertation., 2012

Herstellung in Deutschland:
Schaltungsdienst Lange o.H.G., Berlin
Books on Demand GmbH, Norderstedt
Reha GmbH, Saarbrücken
Amazon Distribution GmbH, Leipzig
ISBN: 978-3-8484-4733-6

Imprint (only for USA, GB)
Bibliographic information published by the Deutsche Nationalbibliothek: The Deutsche Nationalbibliothek lists this publication in the Deutsche Nationalbibliografie; detailed bibliographic data are available in the Internet at http://dnb.d-nb.de.
Any brand names and product names mentioned in this book are subject to trademark, brand or patent protection and are trademarks or registered trademarks of their respective holders. The use of brand names, product names, common names, trade names, product descriptions etc. even without a particular marking in this works is in no way to be construed to mean that such names may be regarded as unrestricted in respect of trademark and brand protection legislation and could thus be used by anyone.

Cover image: www.ingimage.com

Publisher: LAP LAMBERT Academic Publishing GmbH & Co. KG
Heinrich-Böcking-Str. 6-8, 66121 Saarbrücken, Germany
Phone +49 681 3720-310, Fax +49 681 3720-3109
Email: info@lap-publishing.com

Printed in the U.S.A.
Printed in the U.K. by (see last page)
ISBN: 978-3-8484-4733-6

DEDICATION

This research work is dedicated to late Malam Suleiman Yakubu Dutse (*Yaya)* and late Malam Buba Moyole. May their souls rest in peace and may *Jannatul-Firdaus* be their final abode.

ACKNOWLEDGEMENTS

Alhamdu Lillah!

All praises be to Allah, our Creator and Sustainer, the lord of the world; who gave me the opportunity, strength, and determination to complete this research work.

I wish to extend my profound appreciation to my supervisor and co- supervisor Dr. Mustapha Muktar and Dr. Mansur Idris for their encouragements, suggestions, corrections and guide to see this research work becomes reality. May Allah reward them abundantly.

My sincere gratitude and appreciation goes to the M.sc coordinator Associate Professor (Mrs) Binta Tijjani Jibril. May Allah reward her in abundance and ease all her difficulties.

My profound Appreciations to the entire Staff of Economics Department, Bayero University, Kano for their assistance towards the completion of this research study. May Allah reward them with Good deeds.

My sincere gratitude and appreciation goes to Malam Muhammad Inuwa Dutse for his prayers and encouragement. May Allah reward him with *Jannatul Firdaus*.

My sincere appreciation is also extended to the management and staff of Jigawa State College of Education, Gumel, especially my colleagues in the department who contributed immensely to this research project. May Allah reward them in abundance.

My appreciation goes to all my respondents and to all those who contributed directly or indirectly in any way to the completion of this research Study. I have no adequate word to express my heart-felt gratitude to my wife and to my daughter Ikram Balarabe and all my family members for their encouragements and sincere wishes for the successful completion of this work.

Balarabe Inuwa – Dutse
February, 2012.

TABLE OF CONTENTS

CHAPTER ONE

CHAPTER TWO: Literature Review & Theoretical Framework

CHAPTER THREE: Methodology

CHAPTER FOUR: Data Presentation and Analysis

CHAPTER FIVE: Summary, Conclusions, and Recommendation

LIST OF TABLES

LIST OF APPENDICES

CHAPTER ONE
Introduction

1.1 Background to the Study

Food insecurity is said to exist when all people, at all times, do not have physical and economic access to sufficient, safe and nutritious food to meet their dietary needs and food preferences for an active and healthy life (FAO,1996). This definition integrates access to food, availability of food, and the biological utilization of food as well as the stability of all these.

Maxwell and Wiebe cited in Adeyeye, (1997) described Food security as the state of having secure and sustainable access to sufficient food for an active and healthy life. Currently, a synthesis of these definitions with the main emphasis on availability, access and utilization, serves as the working definition in the projects of international organizations.

More than 800 million people throughout the world and particularly in developing countries do not have enough food to meet their basic dietary needs (FAO, 1996). The problems of hunger and food insecurity is a global dimensions and are likely to persist and even increase dramatically in some regions, unless urgent and concerted action is taken, given the anticipated increase in the world's population and the stress on natural resources (Olayemi,1998).

Food security may be analyzed at different conceptual levels: regions, countries, households and individuals. Much analysis of the topic has focused on the macro levels.

Recognizing that the main problem of food security is lack of access rather than an aggregate shortage of supplies, focus on food security has since the World Food Conference of 1974 moved from a global and national perspective to that of household and individual (Robinson, 2001, cited in Shifewa, 2003). Even though food security for individuals is often the main focus of attention, food security is however a measure of a household condition, not that of each individual in the household.

Although food insecurity is closely linked with poverty (Nord, 1999), Therefore, it is incorrect to assume that a state, country, region or municipality poverty prevalence rate is the same as its food insecurity or hunger prevalence rate, since the relationship between poverty and food insecurity is not a consistent one (Osundare, 1999).

Accurate identification and monitoring of food insecurity situations can help public officials, policy makers, service providers and the public at large to access the changing needs for assistance and the effectiveness of existing programmes. While the determination of the food security situation of the households can provide an indispensable tool for assessment and planning. Monitoring food security situation of a particular population may help in comparing the local food security situation to state and national patterns, assess the local need for food assistance or track the effect of changing policies or economic conditions (Bickel, 2000, cited in Omotesho, 2005).

Focus on food security ensures that the basic needs of the poorest and most vulnerable groups are not neglected in policy formulation (WHO, 1993). This is because food security is one of the several necessary conditions for a population to be healthy and well nourished (Nord, 1999). One important aspect of the wealth of a nation is the ability to make food available for the populace. In this connection, food security therefore becomes an important factor in any consideration of sustaining the wealth of the nations (Osundare, 1999). Available statistics show that low average per capita food intake, as well as energy, constitutes perhaps the greatest obstacles to human and national development in Nigeria (Igene, 1997).

1.2 Statement of the Research Problem

Food is a basic necessity of life; its importance is seen in the fact that it is a basic means of sustenance and an adequate food intake, and a key for healthy and active life (FAO, 1996).

Historically, agriculture used to be the mainstay in the Nigerian economy providing food security for the people. It contributed about 85.5 percent of the Nigeria's total export in 1960. However, in 1984, its contribution dropped to 2.6 percent while in 2004, its contribution dropped to as low as 0.81 percent (CBN, 2005).

Nigeria is one of the richest countries on earth in terms of human and material resources. However, despite the oil wealth, vast and fertile agricultural land, Nigeria ranks among the thirteenth poorest countries in the world to the extent that majority of Nigerian populace leave under $1 per day (NBS, 2005).

Poverty is the main cause of hunger and malnutrition, which are aggravated by rapid population growth. The poor are known to have inadequate quantity of food for consumption and the side effect of this is limited growth and brain development (Nord, 1999).

Jigawa State is the 8th most populous State in Nigeria. However, Human Development Indicators of the state are among the grimmest. Even though based on the 2006 CWIQ Survey, only about 47% of all households in the state considered themselves poor (below the national estimates of about 64%), the absolute poverty situation is very high. According to Nigerian Poverty Assessment, the incidence of poverty in Jigawa State is 79.0% which is the highest in the country (NBS, 2012). Severity of poverty in the state put at 24.6% is also among the highest in the country. While there is no state-specific estimate of per capita income, Gross Per Capita Income is below the estimated National Average of N140,000 per annum (SEEDS II, 2009). A number of the socioeconomic characteristics of the population of the state are not favourable. Infant, child and maternal mortality rates are high;

3

access and quality of some of the social services are below acceptable levels. As reported in the Jigawa State Strategic Health Sector Plan (2008 – 2011), mortality rates and the disease burden are 'unacceptably' high.

In Jigawa State as in all other parts of the world, the need for food security has become a policy issue. This need has taken all important dimensions because the basic nutrients in diet (carbohydrates, fats, protein, minerals, vitamins and water) are all necessary for proper body functions. This function includes the production of energy for body activities, the production of growth, the maintenance and repairs of body tissues, the regulation of body processes and the maintenance of proper internal environment. At last poverty is at the corner to prevent Jigawa state indigenes from purchase of food to provide the above functions (Idris, 2009).

However, most of the studies done on this area are mostly on Food Security neglecting Food Insecurity in Nigeria. In line of the above, the researcher finds it necessary to undertake a research on Food Insecurity in order to bridge the gap left by other researchers particularly in the study area.

1.3   Research Questions

The followings questions will be answered by the findings of the research study.

1. What are the factors that determine food insecurity in Jigawa State?

2. What is the possibility of estimating the likelihood of food insecurity in Jigawa State?

3. How does poverty affects the level of food insecurity of household in Jigawa State?

1.4   Objectives of the study

The Major objective of this study is to identify the determinants of food insecurity in Jigawa State, Nigeria. The specific objectives of this study include:

1. To identify the factors that determines food insecurity in the study area.

2. To predict possibility of estimating the likelihood of food insecurity among some selected households in the Study area.

3. To identify the effect of poverty on food insecurity in the study area.

1.4   Justification for The Study

Jigawa state, is blessed with 2.24 million hectares total land area of the state, about 1.6 million hectares is cultivable during the rain-fed season while about 308,000 hectares of the land mass cultivable during the dry season through irrigation (JSEEDS II, 2009).

However, in spite of the above the state remains eleventh poorest in the country with about 79.0% of its population afflicted by poverty as observed by National Bureau for Statistic, hence the justification of the researcher to undertake a study of this nature, in order to identify the determinants of food insecurity and to measure the extent of food insecurity status among the households so as to reduce food insecurity problems in the state by recommending appropriate policies to the government and to all stake holders as well (NBS, 2012).

Research on the Determinants of food insecurity is of paramount importance because it would expose the major causes of food shortage among the communities under the study area.

The significance of this research lies in the fact that it outcome can suggest ways of modernising agriculture in Jigawa State through proper exploitation of its agricultural resources, so as to achieve a desired food security.

The study will also educate Jigawa state policy makers about the major causes of food insecurity in the state and on the need to provide people facing chronic hunger with greater opportunities through job creation. In many state in Nigeria the basic causes underlying food insecurity can only be addressed by accelerating the growth in agriculture. And this depends on whether food policies designed is aimed at increasing the real income and food consumption of the undernourished.

Beside, the research can add to the volume of literatures on the area of food insecurity. This means students, lecturers of tertiary institutions; Agricultural research institutions and other academicians would find the research relevant and very useful.

1.6   Scope and Limitations of the Study

The study does not cover the entire Jigawa State of Nigeria; this is because only six local governments' areas are selected to represent the whole state in identifying the determinants of food insecurity status among the households in the study area.

However, this study was strictly carried out within the frontiers defined by the scope of the study. The study is limited to Six Local Government Areas of Jigawa State which are *Birnin-Kudu, Birniwa, Dutse, Gumel, Hadejia and Ringim,* where two communities each were selected from each local governments and these are; *Kofar Fada & Bakin Kasuwa, Birniwa Town and Birniwa Tasha, Yadi and Fagoji, Dubuntu and Matsaro as well as Sintilmawa and Nassarawa respectively.*

A cross sectional samples of household were taken in December, 2010 with the aim of examining the determinants of food insecurity in the study area.

1.7   Organization of the Study

This research has been designed in five consecutive chapters. Chapter one is made of the introduction, statement of the problem, research questions, and objectives of the study, justification of the study, scope and limitations of the study, organization of the study and definition of abbreviated terms as list of acronyms. Chapter two deals with the literature review, which is divided into conceptual, empirical and theoretical issues.

Chapter three contains the research methodology, and is broken into: Area of the study, methods of data collection, population/sample size of the study, estimation techniques: Descriptive, Econometrics methods, model specification, Data sources and measurements and definition of variables used in the research study. Chapter Four deals with the presentation and analysis of data using both descriptive statistics and inferential analysis.

Finally, chapter five of this research study deals with summary, conclusion and recommendation as well as suggestion for further research studies.

1.9   List of Acronyms

CBN   –   Central Bank of Nigeria

CWIQ   –   Core Welfare Indicators Questionnaires

DFRRI   –   Directorate of Food, Road and Rural Infrastructure

FAO   –   Food and Agricultural Organisation

GDP   –   Gross Domestic Product

ICSFAD   –   American based International Centre for Soil
                Fertility and Agricultural Development.

IFAD   –   International Fund for Agricultural Development

JARDA   –   Jigawa State Agricultural and Rural Development
                Authority

NBS   –   National Bureau for Statistic

NEEDS   –   National Economic Empowerment and Development
                Strategy

RBDA   –   River Basin Development Authority

SEEDS   –   State Economic Empowerment and Development
                Strategy

SPFS   –   Special Programme for Food Security

UN   –   United Nations

USAID   –   United State Agency for International
                Development

WHO   –   World Health Organisation

CHAPTER TWO
Literature Review and Theoretical Frame Work

2.1   The Concept of Food Insecurity

Food insecurity exists when all people at all times do not have access to safe nutritious food to maintain a healthy and active life. Food insecurity or lack of access to nutritionally adequate diet in a household or country can take various forms. For example, chronic food insecurity exists when food supplies are persistently insufficient to supply adequate nutrient for all individuals. Transitory food insecurity occurs when there is a temporary decline in access to adequate food because of instability in food production, food price increases or income shortfalls. We may distinguish between national food security and household food security. This distinction is necessary because an aggregate supply of food, from domestic sources or import or both are prerequisite sufficient income to demand for it (FAO, 1996).

Lack of food excludes people to practice what other people are doing every time. However, large amount of food production in the world does not ensure any country's food security. Moreover, huge production of food at national level does not guarantee for the household food security. This may be due to unfair distribution of resources, variation in production functions, and motives for productivity. That is why even if the production increases through time; food insecurity, malnutrition and hunger would remain the main agenda and much more serious problems in the world today (Fantaw, 2007).

There are two types of food insecurity: chronic and temporal. Chronic food insecurity results from inadequate food intake over a longer period of time and is constant. Temporal food insecurity results from a temporary decrease in food intake due to price changes, production failures, or a loss of income. Temporal food insecurity can also be related to the hungry season.

Food availability is a function of the combination of domestic food stocks, commercial food imports, food aid, and domestic food production, as well as the underlying determinants each of these factors. Use of the term availability is often confusing since it can refer to food supplies available at both household level and at a more aggregate (regional and national) level. However, the term is applied most commonly in reference to food supplies at a regional or national level.

Food access is influenced by the aggregate availability of food through the impact of the latter on supplies in the market and therefore, on market prices.

Food utilization, which is typically reflected in the nutritional status of an individual, is determined by the quantity and quality of dietary intake, general childcare and feeding practices, along with health status and its determinants. Poor infant care and feeding practices, inadequate access to, or the poor quality of, health services are also major determinants of poor health and nutrition. While important for its own sake as it directly influences, human well being, improved food utilization also has feedback effects, through its impact on the health and nutrition of household members, and therefore, on labour productivity and income earning potential.

Food insecurity exists whenever the availability of nutritionally adequate, safe foods or the ability to acquire personally acceptable foods in socially acceptable ways is limited or uncertain. In most populations at any given time, a proportion of people experience some degree of food insecurity because poor are most likely to be hungry in any given country and situation. Around 75 percent of the world's hungry and poor people live in rural areas in poor countries (Campbell, 1991).

Having access to food that is of poor quality or not knowing where the next meal is coming from. Food insecurity can lead to poor physical and mental health, particularly among disadvantaged groups such as the homeless population. Improving food security can therefore improve the health and well being of individuals, communities and entire populations.

Food insecurity may occur at an individual or community level. Factors which indicate risk of community food insecurity include (Wood 2000):

    i)   Absence of a source of free and clean drinking water in local outdoor areas.

    ii)   No local food supply (within 2.5km or walking distance).

    iii)   No car; inability to walk to the nearest food supply; inadequate public transport to the nearest food supply.

    iv)   No real choice in local food outlets.

    v)   Limited choice of food in local outlets (food type, unit size, packaging, quality, and cost).

    vi)   Difficulty buying small amounts of each food required (1-2 servings).

    vii)   Difficulty locating the food wanted inside the nearest food store (poor access and signs).

    viii) Absence of local food outlets that provide a choice of cheap cooked or prepared meals (not just snacks) and a community meeting place.

    ix)   Addictive behavior (prescription drugs, excessive alcohol use, hard drugs, gambling). Wood highlights that the factors that affect food insecurity are the same all over the world, whether a person lives in rural area or urban area.

2.1.1 The Concept of Food Security

Food security is a concept that has evolved during the 1990s far beyond a traditional focus on the supply of food at the national level. This concept has been given general definitions in recent times, but there has been a divergence of ideas on what food security really means (FAO, 1996).

According to (FAO, 1996), food security is defined as access by all people at all times to enough food for an active and healthy life. The committee on world food security defined it as physical and economic access to adequate food by all household members without undue risk of losing the access. However, the definition adopted by the countries attending the world food summit of 1996, and reconfirmed in 2002, accepts the USAID'S concept which has three key elements viz; food availability, food access and food utilization. However, a fourth concept is increasingly becoming accepted namely, "the risks that can disrupt anyone of the first three factors" There are therefore four major elements of food security. They are food availability, food access, food utilization and not loosing such access. Availability, access and utilization are hierarchical in nature. Food availability is necessary but not sufficient for food accessibility and access is necessary but not sufficient for utilization.

In a larger sense, two broad groups of factors determine food security. These are supply side factors and demand side factors. The supply-side factors are those that determine food supply or food availability. In other words, they are determinants of physical access to food at national, household and intra-household levels. The demand side factors on the other hand are factors that determine the degree of access of countries, households and individuals to available food. They are, in other words, determinants of economic access to food or determinants of entitlement to available food. Common to these two sets of factors however is another set of factors that affect the stability of both physical and economic access to foods.

2.1.2 The Concept of Household

The household is the basic unit of analysis in many social, microeconomic and government models. The term refers to all individuals who live in the same dwelling. In economics, a household is a person or a group of people living in the same residence.

United Kingdom, defined household as "one person or a group of people who have the accommodation as their only or main residence and for a group, either   share at least one meal a day or share the living accommodation, that is, a living room or sitting room"

United States defined household as "separate living quarters", i.e. "those in which the occupants live and eat separately from any other persons in the building. Accordingly, Canada defined "a household is generally defined as being composed of a person or group of persons who co-reside in, or occupy a dwelling."

In Sociology 'household work strategy', a term coined by Ray Pahl, is the division of labour between members of a household, whether implicit or the result of explicit decision–making, with the alternatives weighed up in a simplified type of cost-benefit analysis. It is a plan for the relative deployment of household members' time between the three domains of employment:

i)   In the market economy, including home-based self-employment second jobs, in order to obtain money to buy goods and services in the market;

ii)   Domestic production work, such as cultivating a vegetable patch or raising chickens, purely to supply food to the household; and

iii) Domestic consumption work to provide goods and services directly within the household, such as cooking meals, child–care, household repairs, or the manufacture of clothes and gifts. Household work strategies may vary over the life-cycle, as household member's age, or with the economic environment; they may be imposed by one person or be decided collectively.

2.1.3 Historical Perspective on Food Insecurity in the World

Millions of people worldwide suffer from hunger and under nutrition. A major factor contributing to this international problem is food insecurity. This condition exists when people lack sustainable physical or economic access to enough safe, nutritious, and socially acceptable food for a healthy and productive life. Food insecurity may be chronic, seasonal, or temporary, and it may occur at the household, regional, or national level.

The United Nations estimates there are 840 million undernourished people in the world. The majority of undernourished people (799 million) reside in developing countries, most of which are on the continents of Africa (Nigeria Inclusive) and Asia. This figure also includes 11 million people located in developed countries and 30 million people located in countries in transition (e.g., the former Soviet Union). The U.S. Department of Agriculture estimates that nearly 11 percent of U.S. households are food insecure, with approximately one-third of these households experiencing moderate to severe hunger (USDA, 2002).

In developing countries, the root causes of food insecurity include: poverty, war and civil conflict, corruption, national policies that do not promote equal access to food for all, environmental degradation, barriers to trade, insufficient agricultural development, population growth, low levels of education, social and gender inequality, poor health status, cultural insensitivity, and natural disasters. In the United States, the primary cause of food insecurity is poverty. Low levels of education, poor health status, and certain disabilities also increase the risk of food insecurity for individuals and households in the United States (Nord, 2002).

Globally, certain groups of people are more vulnerable to food insecurity than others. Vulnerable groups include: victims of conflict (e.g., refugees and internally

14

displaced people); migrant workers; marginal populations (e.g., school dropouts, unemployed people, homeless people, and orphans); dependent populations (e.g., elderly people, children under five, and disabled and ill people); women of reproductive age; ethnic minorities; and low literacy households (FAO, 2002).

Food insecurity affects millions of people around the world, including these children in Thailand. The situation in that country and a handful of others has improved slightly, but progress is slow.

However, enough safe and nutritious food is either domestically produced or imported from the international market. However, food availability does not ensure food accessibility. Government policies must also contribute to equal distribution of food within nations, regions, and communities. In addition, for food to be accessible, individuals and families must be able to afford the food prices on the market. Finally, food must be properly utilized. Proper utilization depends on proper food storage to guard against spoilage, appropriate handling to avoid disease transmission, and proper preparation to ensure nutritiously balanced meals.

Individuals need adequate amounts of a variety of quality, safe foods to be healthy and well-nourished. Under nutrition results from an insufficient intake or is an improper balance of protein, energy, and micronutrients. Nutritional consequences of insufficient food or under nutrition include protein energy malnutrition, anemia, vitamin A deficiency, iodine deficiency, and iron deficiency.

Food insecurity and malnutrition result in catastrophic amounts of human suffering. The World Health Organization estimates that approximately 60 percent of all childhood deaths in the developing world are associated with chronic hunger and malnutrition. In developing countries, persistent malnutrition leaves children weak, vulnerable, and less able to fight such common childhood illnesses as diarrhea, acute respiratory infections, malaria, and measles. Children who are mildly to moderately malnourished food are at greater risk of dying from these common diseases.

Malnourished children in the United States suffer from poorer health status, compromised immune systems, and higher rates of illnesses such as colds, headaches, and fatigue. (WHO, 2002)

Adolescents and adults also suffer adverse consequences of food insecurity and malnutrition. Malnutrition can lead to decreased energy levels, delayed maturation, growth failure, impaired cognitive ability, diminished capacity to learn, decreased ability to resist infections and illnesses, shortened life expectancy, increased maternal mortality, and low birth weight.

Food insecurity may also result in severe social, psychological, and behavioral consequences. Food-insecure individuals may manifest feelings of alienation, powerlessness, stress, and anxiety, and they may experience reduced productivity, reduced work and school performance, and reduced income earnings. Household dynamics may become disrupted because of a preoccupation with obtaining food, which may lead to anger, pessimism, and irritability. Adverse consequences for children include: higher levels of aggressive or destructive behavior, hyperactivity, anxiety, difficulty with social interactions (e.g., more withdrawn or socially disruptive), increased passivity, poorer overall school performance, increased school absences, and a greater need for mental health care services (e.g., for depression or suicidal behaviors).

To understand the magnitude of food insecurity, hunger, and malnutrition, one must consider both the continued rapid growth in world population and the number of individuals below the poverty line. In 1999 the world population reached 6 billion. The United Nations estimates the world population will exceed 8 billion by 2025. In terms of poverty, the World Bank estimates that nearly 1.2 billion people live on less than one dollar a day, which is the internationally recognized standard for measuring poverty. Another 2.8 billion live on less than two dollars a day. (USDA, 2002)

In addition to these progress-slowing conditions, the number of under-nourished people is actually growing in most developing regions. A few large countries have made significant gains, making the global picture appear more promising than it really is. China, Indonesia, Vietnam, Thailand, Nigeria, Ghana, and Peru have all made important gains in reducing food insecurity and hunger. However, in nearly fifty other countries, the number of undernourished people increased by almost 100 million between 1993 and 2003. The absolute numbers continue to rise as a result of rapid population growth, even though the proportion of undernourished people in most developing countries is actually decreasing. (WBG, 2002)

Worldwide commitment to improve global food insecurity was demonstrated at the 1996 World Food Summit, where 186 countries pledged to reduce the number of hungry, food-insecure people in the world by 50 percent (to 400 million) by the year 2015. Progress toward this goal has been slow, with a decrease of only 2.5 million people a year since 1992. At the current pace, the goal will be reached more than one hundred years late. Despite slow progress, some innovative programs have been implemented around the globe to combat food insecurity and under nutrition. Examples of innovative program include: community gardens, farmers markets, community-supported sustainable agricultural programs, and sand food for work exchange programs, farm to school initiatives, credit to poor households, income transfer schemes, and agricultural diversification programs. ((FAOUN, 2001)

Food insecurity remains a significant international problem, with developing regions of the world enduring most of the burden. Food insecurity results in considerable health, social, psychological, and behavioral consequences and is undeniably linked to poverty. Despite international commitment, the number of food insecure individuals remains unacceptably high.

2.1.4 The Magnitude of Food Insecurity in Africa

Although there is considerable variation across countries in this region, the regional statistics and trends presented below paint a stark portrait of food insecurity.

In the late 1980s, an estimated 71 million people, or 46 percent of the region's population, were chronically food insecure. This percentage is higher than that for sub-Saharan Africa as a whole and has continued to increase in the past five years.

In August 1994, approximately 22 million people in the region were in need of external food aid assistance. Nearly 11 million of these were refugees and internally displaced people, most of whom fled their homes as a consequence of civil strife. Another 11 million are severely drought- affected. Most refugees and internally displaced people do not have the resources for sustained access to food supplies. The 7.3 million internally displaced were concentrated in five countries: Sudan, Burundi, Kenya, Rwanda and Somalia. Every country in the region except Eritrea hosts some of the 3.8 million refugees. Per capita food production declined in the region by more than 16 percent over the 1980 to 1993 period (World Bank, 2000).

As domestic food production per capita has declined, food import bills have placed increasing strains on trade balances and donor food aid assistance has increased substantially. The value of food imports as a percentage of export earnings has risen from a low of about 27 percent in 1980 to more than 35 percent in the early 1990s. The capacity of most countries in the region to significantly increase commercial imports is limited by low export earnings.

Regional per capita calorie availability (1,950 Kcal per capita per day) is less than the international minimum standard for survival of 2,100 calories and much less than the standard for an adequate diet of 2,400 calories. These current levels, stagnant for the last 10 years, are even below those attained in the region in the 1960s (USAID, 1999).

18

The Greater Horn region is one of the poorest in the world. Gross National Product per capita is US$167, and growth rates, which were negative throughout most of the 1980s, averaged -2 percent in the early 1990s. Poverty analyses done by the World Bank suggest that skewed income distribution in some countries (Kenya, Tanzania) makes food access a struggle for many even when food is nationally available (WBG, 2002).

Poor nutritional and health status indicators are another dimension of high food insecurity. The regional infant mortality rate is 107 deaths per 1,000 with malnutrition underlying more than one-third of infant and child deaths. The prevalence of wasting (low weight to height measurement) of children under 2 years was between 30 percent and 45 percent in 1990 for the six countries for which data were available. Malnutrition also may be implicated in up to 80 percent of maternal deaths.

 The magnitude of food insecurity in the Greater Horn of Africa remains a compelling human crisis.

2.1.5 The Root Causes of Food Insecurity in Africa

At the root of this alarming description of food insecurity is an unstable social and political environment that has precluded sustainable economic growth. A number of factors have converged to create this instability: Poor economic policies have inhibited the development of agriculture based on comparative advantage and intensification of agriculture, retarding economic growth; growing population pressures have combined with a lack of investment in human resource development, further stressing the natural resource base; civil strife and a scarcity of democratic institutions have undermined sustainable growth strategies; and the natural resource base of the region is highly uneven, and several countries have limited areas of high agricultural production potentials. Linked to weak national institutions are weak regional institutions precluding effective action on these underlying causes. These causes and their relative importance should be jointly analyzed with African organizations to help guide integrated efforts to overcome food insecurity.

1.   Natural Resource Constraints

The limitations of the natural environment in the Greater Horn place certain constraints on improving food security. The chances of drought occurring in parts of the Greater Horn have increased from a probability of one in every six years to one in three years for those areas affected. Over the last decade there are two apparent changes in long-term weather patterns. First, there is a mean decrease in annual rainfall in the Sahelian Zone of Sudan; and second, inter annual variability of rainfall has been increasing in the crescent from Kenya to Sudan, including parts of Ethiopia and Tanzania.

Repeated occurrences of drought and high variability in precipitation have reduced the ability of many smallholders to maintain their assets or to respond when conditions are good. Other natural disasters such as pest infestations and periodic flooding destroy area-specific production levels. Analysis of these factors argues for a more effective weather and pest early warning system in the region.

Increases in agricultural production in the past in many of the Greater Horn countries resulted from expansion onto new lands; notable exceptions include parts of Kenya, Rwanda and Uganda, where land scarcity has led to intensified use of existing land. The movement onto new lands, without improvement of management techniques, has caused environmental deterioration. In several countries, there is little additional land to be brought into production -- but there exists potential for agricultural intensification. However, in many areas few incentives exist to intensify. Roads are inadequate, and market distortions are common due to poor policies, including pricing, land rights, and, in some areas, poorly targeted food aid programs. Arable land in Ethiopia exists, but required investments in infrastructure and malaria eradication would be costly. Unexploited water resources exist, but the potential has not been fully explored and can be costly. The current accounting shows that our knowledge of the region's natural resource endowment is less than adequate.

Lack of agricultural intensification and low agricultural productivity mean that many of those in rural areas remain subsistence producers, and, therefore, the large quantity of food at low prices which is essential for economic growth in urban areas is not available. Productivity increases and growth linkages both in rural areas and between urban and rural areas are a prerequisite to increased incomes and sustained access to basic foodstuffs. Other food sources that have been neglected and could provide these linkages include livestock and aquatic resources. Aquaculture could be enhanced, and the rational use of marine resources could be promoted (six of the 10 countries have access to marine resources). Livestock are also extremely important throughout the region, but continued difficulties in policies such as disease control, grazing rights and supply of important inputs inhibit production increases. A component of a successful regional food security strategy needs to focus on regional research and diversification in resource management, and growth linkages throughout these economies.

2. Poor Economic Policy Environment

Agricultural intensification and the development of product markets and processing industries have not occurred partly because of a poor policy framework that has led to inadequate research, a lack of appropriate technologies and weak dissemination of existing technologies. Many policies persist that neglect the critical role of women in agriculture and restrict their full involvement in that and other sectors.

Economic and agricultural policies that distort prices of agricultural inputs and outputs adversely affect investment in agricultural production, marketing and storage, and incomes earned from these activities. While structural adjustment programs have improved some critical pricing and other administrative control issues, difficult policy problems remain, especially privatization, land tenure, access to critical inputs, and international and domestic trade. Policy reform is a high priority because it is a necessary, though not sufficient, condition for growth.

Weak market integration due to both poor infrastructure and domestic and international trade restrictions affect the ability of food surplus countries and regions to export to food deficit countries and regions. Public infrastructure, an important condition for both food availability and access, remains limited in all countries. For example, road systems do not reach the majority of the population in Ethiopia and are a woefully inadequate support to an active and extensive agricultural trade in Tanzania. The eight major railroad systems in the region are in various stages of disrepair. A Cross-border trade restriction as well as restrictions on internal movement of foodstuffs has prohibited the private sector from responding to shortages when and where they exist. Some progress has been made in lifting these restrictions, but additional policy analysis and reforms are needed.

Liberalization of marketing systems has encouraged the growth of small trading firms and entrepreneurs in several countries, but viable commercial enterprises throughout the food system (storage, assembly, processing and marketing) are lacking. The lack of development of efficient services is to some extent linked to the inadequacy of roads and the availability and cost of trucking. In many areas private traders are not able to respond to the liberalized markets because of lack of access to working capital. In addition to policy constraints, firms in this area need technology, financing and management support to increase food availability at low cost.

3.   Rapid Population Growth and a Poor Human Resource Base

Population growth rates in the region were very low before 1950. In the four decades from 1954 to 1994, population tripled, growing from about 61 million to approximately 186 million. The current regional population growth rate is 2.9 percent. A partial explanation for food insecurity is that the greatly increased population of the Greater Horn may have approached or exceeded the carrying capacity of the fragile environment in some areas. With reduced fallow, barely arable land being cultivated and increased grazing pressures to feed livestock, increasing soil erosion and deforestation are reducing productivity. High population growth must be dealt with immediately, even though the impact of policies implemented now will only occur over the long-term (World Bank, 2002).

Population growth rates remain high because of poverty and accompanying problems of poor child survival rates. Limited access to or the absence of basic services such as health facilities and education has meant low contraceptive prevalence rates, poor bodily absorption of available food nutrients because of disease and high infant mortality. Illiteracy rates in the region, especially high for women, exceed the average for sub-Saharan Africa as a whole and correlate closely with fertility and high infant mortality rates. Another urgent problem that transcends borders is the spread of HIV/AIDS throughout these countries.

These poor levels of health undermine increases in labor productivity and represent a considerable barrier to increases in growth. Improvements in health services and education, especially those targeted to women, will contribute to reducing population growth rates and average family size over time. A healthy and educated population will, in turn, contribute to productivity increases and economic growth.

4.   Civil Strife and Absence of Good Governance

For the last 25 years, armed conflict has been endemic to the region; since 1980, the number of Greater Horn countries suffering the consequences of civil strife at any one time has increased from three to five or six per year. Both military confrontation and attendant political instability have undermined food security in the region in direct and indirect ways. Agricultural production has been disrupted by actual conflict, by war-induced mass migrations and by an emphasis on defense over and above agricultural and social sector spending. Each of the greatest food crises of recent years -- Ethiopia in 1984-85, Somalia in 1992, Sudan since 1987 and Rwanda in 1994 -- has been generated by conflict.

Insecurity in the Greater Horn region ranges from full-scale warfare to cross-border clashes. In all cases, the militarization of the region during the Cold War era has contributed to the scale of conflict as well as to the tendency to resolve disputes by force. Governance in the Greater Horn region has for decades been characterized by autocratic rule and by extensive centralization required to maintain state power. This has, in turn, resulted in the growing political and economic disenfranchisement of the majority and in the retardation of the growth of democratic institutions in government or civil society. Armed conflict has evolved as the most readily available course of political interaction.

The structures of governance in the region have not only arrested popular participation in political decision-making and spawned armed conflict; they also have directly affected development. Weak, unstable or highly centralized governance structures have proven inadequate to the task of responding to food crises, even in cases where early warning indicators might have allowed for famine mitigation or prevention. The militarization and centralization of governance also

24

has meant that the social sectors, including, for example, education, have received low priority. Illiteracy rates for the region are among the highest in Africa. Total illiteracy is above 50 percent for all countries but two, and in three countries, illiteracy rates for women exceed 85 percent. A healthy, well-trained labor force is a prerequisite for increased economic production; and at the same time, education is both a means and an end to a responsible citizenry.

There are cases for both optimism and pessimism in the potential for resolution of these crises in the region. In some parts of the Greater Horn the tide has turned. The 30-year war between Ethiopia and Eritrea has been resolved. Uganda has emerged from a past of terror and armed conflict. However, the Greater Horn region has seen -- in Somalia and Rwanda -- the emergence of "failed states".

5.   Weak Regional Institutions and Donor Coordination

The absence of stable and legitimate national governance structures and the continuation of cross-border conflict have precluded strong regional organizations to deal with complex regional problems such as refugees, trade, arms flows, natural disasters, etc. Regional analysis and action are needed in these areas, and forums could be supported in which comparative experiences can be shared to increase options. One regional institution, the Inter-Governmental Authority on Drought and Development, exists but is relatively weak. However, African leaders have indicated their support to strengthen IGADD, and other donors have initiated actions in this direction.

Donor involvement in the region has often been duplicative, conflicting or conducted without local participation. No mechanism comparable to the Club du Sahel exists that could improve collaboration among donors, and integrate actions of donors and African organizations. In-country donor coordination is often limited to the exchange of information after implementation. Not only is a regional coordination forum lacking, but no national structure exists for joint problem identification and joint resolution with donors and non-governmental organizations (NGOs). Societies of the region lack recognition of the benefits that accrue to nations from regional coordination. National organizations need to move beyond their preoccupation with national-level solutions where efficient and effective regional approaches can be formed.

2.1.6 History of Food Insecurity in Nigeria

Food insecurity problem in Nigeria can be traced back to colonial period where a Nigerian peasant communities were producing a variety of commodities to satisfy their needs and the remaining little surpluses were used in exchange with other commodities hence Agriculture then involves production of Food crop for subsistence (Kwanashie, 2010).

Food insecurity problems in Nigeria began with the introduction of tax system by colonial masters making Nigerian peasant farmers to divert their mode of production from food crops to cash crops in order to pay taxes to colonial masters while the proceeds are exported to the western world.

However, before the coming of the white men into Nigeria, people doest have much food insecurity problem since they can afford three square meals. However, at that period Agriculture was the major source of employment in Nigeria employing more than 80% of the rural population and is also the major source of export earnings to the Nigerian Economy.

Food Insecurity problem become worse in Nigeria due to the less priority given to the Agricultural sector due to the discovery of oil in the sixties. For example, in 1960 the contribution of Agriculture to the Nigerian GDP is 85.5% but latter the contribution dropped to 2.6% in 1984 and subsequently in 2004 the contribution dropped to as low as 0.81% (CBN, 2005).

Food Insecurity in Nigeria become pervasive this is because food importation of Nigeria also increases at increasing rate up till today. For instance in 1980 the total Food Import to Nigeria stood at N1, 437.5 Million, to N103, 289.8 Million and N290, 654.9 Million in 1999 and 2007 respectively (CBN, 2007).

2.1.7 Factors Responsible For Food Insecurity

Production, income, and the high price of food are the variables that contribute to hunger in rural areas. In terms of production, poor farmers only make 10-20 percent of what is produced on green-revolution farms. In terms of income in relation to the price of food, in less developed countries (LDCs) 50 to 80 percent of income goes to food. In more developed countries (MDCs) the percent is 10 to 15 (FAO, 2005). The poor are vulnerable to changes in prices, income, and wellbeing; and most dependent on social services such as health and welfare services. A small change in any of these variables can easily push a family into hunger, further decreasing their ability to procure food and income, and a cycle of poverty and hunger ensues.

Poverty leads to hunger and vice-versa; families caught in a cycle of hunger and poverty find their opportunities and resources further diminished in other areas. The importance of fulfilling all the MDG goals illustrates this point well. The 2005 State of Food Insecurity in the World report argues that without rapid progress in reducing hunger, achieving all of the other MDGs will be difficult or impossible. Hunger and malnutrition lead to poverty, which leads to: a.) Unsustainable use of natural resources; b.) Reduced capacity to access markets and resources; c.) Reduced school attendance and learning capacity; d.) Less education and employment for women and girls; e.) Weakened immune systems and rising child mortality; f.) Impaired maternal and infant health; g.) Risky survival strategies, spread of HIV/AIDS, malaria and other diseases; and h.) Unsustainable use of natural resources.

The individuals within food insecure households cannot be assumed to suffer from hunger equally; there are differences in distribution and negotiating abilities of

27

individuals. There are generally inequalities in regards to age, gender, and disabilities. For example, within the household, women and children are more likely to suffer from food insecurity because of their limited control over resources; men are more likely to eat first in the house as the producers of food and income. It has been repeatedly demonstrated that female-headed households are more likely to suffer from food insecurity because the lack access to resources, education, land, and are less likely to find work (FAO, 2005, cited in Idris, 2009).

2.1.8 Poverty, Economic Growth and food insecurity

Poverty is about lack of opportunity, not just lack of income. The lesson from the last 50 years economic growth is the most powerful way of pulling people out of poverty. Economic growth creates higher income, which helps people to save, invest of protect themselves when times are hard. Higher family income means children can have access to good adequate nutrition and education rather than have to work. However, as economy grows government can raise the money for public service.

Over recent decades, Asia has seen dramatic economic growth, first in the eastern Asian Tigers of South Korea, Singapore and Taiwan, then Thailand, Malaysia and Indonesia and more recently China and Vietnam. Trade and openness to the international economy has been the key to this economic success. There has also been significant progress is South Asia. In 1990s, economic growth helped reduce poverty in the region from just over 40% to around 30% (World Bank, 2006).

The picture is very different in sub-Saharan Africa, though there have been some success stories. Uganda and Ghana for instance, had high enough growth during the 1990s to reduce poverty by more than 10%. But the percentage of people living in poverty in the region as a whole has increased in the last two decades. There are now over three million poor people in sub- Saharan Africa (World Bank, 2006).

28

Poverty and stunted growth have constituted two major challenges that Nigeria has faced. The trend in relative poverty has been an increasing one with the percentage of the population in poverty rising from 46.3% in 1982 to 54.4% in 2004. The National bureau of statistics estimate for 2004 also shows that more than half of the population lived below the international bench mark of $1 per day, an estimate that is even considered conservative (Taiwo, 2008). In 2004, more than three – quarter of Nigerians were considered poor and unsecured.

Nigerian remains a low income and low growth country. During 1990- 2006, for instance, the Gross Domestic Product (GDP) rose from N326.6 billion to 593.97 billion at an average rate of 3.8%. Growth performance received a fillip during 2002 – 2006, with GDP growth rate ranging from 4.6% the following year. The growth of per capita income was less imperative. Per capita income decline in the 1990sw but it has been growing since 2000. On average, during 1990-2006, per capita income grew by just 0.9% (CBN, 2005, cited in Idris, 2009)).

The incidence of poverty varies markedly between states. For instance in 2007 Jigawa record the highest followed by Kebbi, Kogi, Bauchi and Kwara States. Food insecurity is an important index of poverty. Households are food secure when they have year- round access to quality and quantity of foods their members need for an active and healthy lives. Food shortages when ever occurs leads to anxiety and even national insecurity. In addition, adequate food availability and security is important for health, national productivity, mental and physical developments of citizen. Similarly price stability, is a major factor facilitating access to food in Nigeria. Alas, food insecurity can only be alleviated when poverty is alleviated from the general public (MDG's Report, 2005 as cited in Idris, 2009).

2.1.9 Socio Demographic Characteristics of Households and Food Desire

Most scholars consider consumption pattern and "food ways" interchangeably. Schiff and Valdes (1990) described food wants as a societal development traditional ways of conducting the activities connected with food. That is, how food is acquired, how it is prepared for eating, with whom, when and in what quantity Schneff (19992), describe food ways as follows "man does not want food or drink as biologically defined. He wants a particular kind of food and drink, cooked or prepared according to his taste, serve in a manner and quantity appropriate".

Generally, two factors serve as determinants of food desires. The most important factor affecting food consumption pattern is the availability of food supply and its acceptance. Culturally define values, attitudes and believes always form the frameworks within which food consumption pattern develop. Others are safety and health benefits and conditional responses to food.

Although in analyzing household expenditure behavior emphasis has been placed on expenditure income (Engel) relationship. However, Cheng and Capps, (1998), opined that all household may not face the same prices and further, preferences for the commodity in question may not be the same across households. Thus, the effect of household income, price and socio – demographic factors on expenditure should be considered simultaneously.

Bourguignon *etal* (1998), reported the distribution and consumption of food within the nuclear family symbolized relationships. They also identified educational level of household's heads as a factor affecting pattern of food consumption.

However, Jha *etal* (2000), opined that demographic characteristic and life style recognition that reforms may not directly affect security, rather the impacts are indirect. The impact of agricultural prices, production and trade through changes in relative prices constitute the intermediate impacts. Finally, there are the impact on the food security status of individual households and the nation at large. In order words, the reforms are expected to have some impact on the price, production and the trade before affecting food security indicators, cited in Idris 2009).

2.1.10 Agricultural Policies in Nigeria

Since independence various governments have made several efforts to do away with food insecurity problem in the country. In the 60s Nigeria depended on agriculture to provide infrastructure and run services until the collapse of the first republic, and the military takeover of government in 1966. At that time a lot of seedlings were taken to other countries. For example, palm fruits were exported to Malaysia. Nigeria also excels in the production of Cocoa, rubber, groundnut and cotton. Subsequent administration saw the need to reverse the downward trend after the first republic.

In the 70s the Government introduced the National Accelerated Food Production programme and the Nigerian Agriculture and Cooperative Bank was established to fund agriculture and assist farmers. This was followed by Operation Feed the Nation in 1976. The programme was fashioned to revolutionise agricultural sector of Nigerian economy, which was derailing from its normal contribution to the economy. For instance, between 1965 and 1970 the percentage share of agriculture in total GDP was 54.8. This dropped to 38.6 percent between 1971 and 1975 and reduced further to 21.1 percent in 1976 to 1980. To make the programme effective, farmers most especially in the rural areas were taught farming practices and agriculture was made compulsory in all secondary schools.

In addition, eleven River Basin Development Authorities (R.B.D.A.), were established to facilitate irrigation agriculture as an attempt to expand farmland. Also, farm settlements were established for cash and food crops to reduce food importation and ensure self sufficiency in food production.

Government's efforts between 1981 and 1985 yielded good fruit as the contribution of agriculture to GDP rose from 21.1% to 35.4%. This was the result of the implementation of Green Revolution Programme of Shehu Shagari Administration which complements the R.B.D.A Programme.

The Military Administration of General Badamasi Babangida in 1986 came up with rural infrastructure development programme and established the Directorate of Food, Rood and Rural Infrastructure (DFRRI). The programme was to open up rural areas for effective agricultural activities and boost food production. The effort raised the contribution by agriculture to 39.9%. The trend changed since 1991 and the contribution has been decreasing, and Nigeria has been depending more on food importation. This was the period of essential commodity' and the beginning of high level of corruption in the country. People collect money for contracts which were not executed.

The Democratic government of President Olusegun Obasanjo has so many policies and programmes; reorganizing, restructuring, privatizing institutions and agencies was in partnership with some others to make impact. The programs, policy instruments, and policy direction enumerated above are clear indications of government's interest in and commitment to, increased food production. According to the government (NEEDS, 2004), the numerous initiatives are expected, ceteris paribus, to:

Provide incentives for private sector participation in the agricultural sector, foster effective linkage with the industrial sector; add value to agricultural produce through processing for export; create more agricultural and rural employment opportunities; increase the income of farmers; reduce drastically the rising trend in food import and ultimately achieve food security.

All these have positive impact on agricultural production and consequent improvement in the contribution of agriculture to total gross domestic product. For instance, a sharp increase in contribution from 24.6 percent between 1996 and 2000 as against 42.20 percent in 2007. Despite all these efforts, not less than 65 percent of Nigerians are food insecure (Mohammed, 2008).

However, various governments have made efforts in revamping agricultural practice to minimise food insecurity problems in Nigeria but all effort in vain. In addition, if proper care and attention will be given to the Agricultural secretor alone by the Government, Agriculture will still remains the mainstay of Nigerian Economy.

2.1.11 Major Problems Facing Agriculture in Nigeria

Okunye, (19997) cited in Idris (2009), noted that agriculture has been beset by long standing problems impending its productivity and contribution to national aggregate output. According to him these problems can be grouped into the followings:-

1. Infrastructural facilities: which includes
   a) Poor feeder roads and inadequate road network between the rural areas where agricultural production.
   b) The rural electrification programmes in Nigeria have not fully taken off as the government battles with the supply of electricity in urban areas.
   c) Irrigation facilities are still very poor despite the existence of River Basin and Rural development Authorities (RBDA).
   d) Schools (Primary and Secondary) are few in the Rural Areas and hence the migration of youth to the urban areas becomes imperative.

2. Manpower/Skill Development: which includes,
   a) The extension service delivery system still suffers from inadequate number of extension workers. The few ones that are in place, lacks mobility to improve on the extension farmer contact while women extension are few to handle gender issues.
   b) The frequency in extension massage discovery is limited by poor research situations in Universities and Research Institutes.

c) Shortage of experienced professionals and technical manpower especially for tractorisation and mechanization, generally.

2.1.12 Agriculture and Food Security in Jigawa State

Agriculture is the mainstay of Jigawa State's economy as it provides livelihood for close to 90% of the State's population. Despite the existence of high potentials for market-oriented agricultural production farming in the State is mainly for subsistence. Jigawa is a rural and agrarian state where majority of the people earn their living through subsistence farming that relies heavily on rainfall using traditional implements. The major crops produced in the state include millet, groundnut, corn, and maize. Many farmers also engage in rearing of livestock, such as cattle, goats and sheep. Despite increased land fragmentation that has taken place over the years, majority of the population have access to farming land with an average family plot measuring approximately between 2.5 and 3.0 hectares (JARDA, 2009, cited in JSEEDSII, 2009).

The State is blessed with large expanse of agricultural land, rivers and flood plains suitable for crops, livestock and fish production. Out of the 2.24 million hectares total land area of the State, about 1.6 million hectares are estimated to be cultivatable during the rain-fed season while about 308,000 hectares of the land mass is cultivable during the dry season through irrigation. Based on this, over 80% of the State's total landmass is considered arable, which makes it one of the most agriculturally endowed States in Nigeria. This arable land comprises of: -

a) Upland soils – which are characterized by low organic and nutrient Content. It is largely used for rain season farming with potentials for Irrigation farming including the development of orchards;

b) Fadama soils – which are of higher organic and nutrient content regularly replenished by seasonal flooding. The Fadama flood plains (about 150,000 ha) are rich in both surface and subsurface water, which makes it

34

Amenable to both rain-fed and irrigation farming. Jigawa State is endowed with abundant livestock resources. Popular livestock species in the State include goats, sheep, poultry and cattle with estimated population of 2.3 million, 1.8 million, 4.2 million and 1.1 million respectively (JARDA 1993/94 LSR). The state also has an estimated 450,452 square kilometers of grazing land reserve, which provides opportunities for large-scale and sustainable livestock development. Currently there are twenty major irrigation schemes and fourteen borehole-based irrigation schemes in the state. Some of these irrigation schemes are located in the Fadama areas. The Hadejia Valley Project, which consists of a vast expanse of irrigable fertile Fadama land covering over 4,800 hectares, is perhaps the greatest single agric potential of the State.

Water is one of the most precious resources of Jigawa State with the State sitting atop two important aquifers of significant ground water resources. Surface water storage capacity totals approximately 477 million cubic meters. Ground water potential is significant with Jigawa State falling in to the pumping range of 30,000 – 40,000m3/km2. A total of approximately 3,676 million cubic meters per year are recharged to the ground and surface water storage from rainfall as calculated using Water Balance Analysis. Agric growth is considered a catalyst for economic empowerment to achieve sustained poverty reduction and food security (SEEDS II, 2007).

The first MDG is to eradicate extreme poverty and hunger amongst the population. For Jigawa State which has over 85% of its population residing in rural areas and engaged in subsistence agriculture achieving this goal without sustained growth in the agricultural sector would be difficult. As a leading sector providing employment for the majority of the people Agriculture is an obvious choice in Government's poverty alleviation and economic empowerment strategy.

Empirical evidence shows that there is a positive relationship between growth in the agricultural sector and poverty reduction. According to the World Bank Rural Development Strategy dubbed "Reaching the Rural Poor", a 10% increase in crop yields leads to a 9% reduction in the number of people living on less than US$1 a day. To ensure that the adopted strategy is sustainable, agriculture is treated along with environment as a single pillar. Apart from its pot ntials for achieving food security and employment generation, the sector is also critical in providing the necessary forward and backward linkages in the development of small scale enterprises.

2.1.13 Policy Objectives of Agriculture in Jigawa State

Consistent with the Millennium Development Goals, the key objective of the agric sector is to achieve substantial poverty reduction, increased food security and nutritional value especially for women and children and contribute to sustainable employment opportunities through sustained agric growth and economic empowerment of the farmers. Another strategic objective is for the agricultural sector to play a prominent role in providing an enabling environment for Investment and agro-based economic growth. These policy objectives take into account the high incidence of rural poverty and the predominance of the rural poor in the state's population. However, the following targets are made to achieve;

Due to absence of baseline data it is not possible to know the current performance of agric production in the state. The agric targets stated below are tentative and are expected to be corrected when data becomes available:

i)   Increased agricultural productivity (higher farm yields) by 10% during the period 2009 to 2011.

ii)   Reduce post-harvest losses by 25% for both cereals and vegetables by 2011;

iii)   Increase profitability of agricultural production with 20-30% by year 2011.

However, in order to achieve the objectives and targets outlined above, it would be necessary to pursue a multi-faceted strategy in which the various aspects of the agriculture sector are brought together to achieve desired results. Two critical areas are pre-harvest input delivery services (including supply of seeds, fertilizer, and research and extension services) and post-harvest support with respect to storage, processing and marketing of agric produce covering both food crops and industrial crops. The strategy would also include a rural development component that target the provision of rural infrastructure for sustainable rural livelihoods. Some of the specific measures that would be instituted and pursued under the agric-sector strategy are as follows:

i)   Supporting the cultivation of off season crops which constitute part of the livelihoods of many households in the state in both Fadama and non-Fadama areas; This also entails the promotion of irrigation agriculture involving development of small-scale irrigation development through the provision of 10,000 shallow tube wells fitted with two-inch petrol driven water pumps in the next three years; rehabilitation of existing irrigation schemes numbering 29 (Including 14 borehole-based schemes) across the state to facilitate the cultivation of about 4,000 hectares of irrigated land in the next 3 years; and establishment of 20 additional direct pumping schemes of 120 Ha along River Hadejia in the next three years;

ii)   Ensuring timely provision and access to agriculture inputs including improved seed varieties, fertilizer, insecticides and modern farm implements. This would entail support to the Jigawa State Agricultural Supply Company (JASCO). As part of the strategies for ensuring supply of high quality seeds to the farmers, a seed processing plant would be established in the state by year 2011. Seeds to be promoted include millet, corn, sesame, rice, maize, cowpea and vegetables, amongst others.

iii)   Active support for livestock development through development and protection of additional 30 grazing reserves, revitalization of existing Livestock Improvement and Breeding Centres and Ranches located at B/Kudu, Gumel, Birniwa and Kazaure in the next 3 years and promotion of viable local livestock breeds including small ruminants;

iv)   Promoting agricultural research and extension services particularly in the areas of adaptive research and technologies, soil fertility, integrated pest control, reduction of post-harvest losses etc. This is to be achieved through

38

support to the State Agricultural Research Institute and agricultural extension agents of the State ADP;

v) Promotion of animal traction with emphasis to patronizing local fabricators of ploughs through the disbursement of additional 1,500 packages of work bull, ox cart and plough by year 2010;

Government will support and collaborate with donor agencies such as (IFAD and FADAMA DEV) in funding special projects. This will focus on:

a) Monitoring and control of pests and disease outbreaks on crops and livestock and carrying out routine vaccination exercise on livestock;

b) Improve access to agricultural credit by 30% through linkage with and support to financial institutions with emphasis on value addition and processing activities done by women.

c) Gradual modernization / mechanization of production techniques to reduce drudgery and increase labour productivity in both on-farm and off-farm production and processing activities. This would entail the scaling up of the work bull animal traction programme, promoting and facilitating the use of modern agricultural tools and implements such as tractors, harvesters and thrashers. Some of these would require active collaboration with various stakeholders in the public and private sector including support in the area of agricultural marketing;

d) Review of the state grains buffer stock programme to ensure its optimal utilisation as a programme of strategic food reserve. This would also entail monitoring of markets trends to ensure timely interventions to stabilise prices and ensure availability and accessibility to food by all and at all times;

e) Development of artisanal and culture fish production by 30% in the next 3 years Overall the agric policy and strategy seeks to take advantage of existing potentials and to mitigate identified constraints in order to achieve increased agricultural production and food security in the state.

2.1.14 Major Problems of Agriculture in Jigawa State

A major constraint to agricultural production and food security is income poverty which hampers the peoples' access either to the basic agricultural inputs as farmers or limits their purchasing power in buying food. Other major impediments to the growth of agriculture in the state include:

i) Subsistence smallholder system, which is an inherently low-productivity system Characterized by low farmers' capacity to apply improved farming practices.

ii) Low nutrients content of the soil which is exacerbated by low productivity of crops varieties and livestock species;

iii) Periodic pests and diseases out-break including Quela birds and grass hoppers.

iv) High post-harvest losses due to poor storage and absence of cost-effective small and medium scale processing equipment as well as inadequate or ineffective distribution and marketing systems;

v) Drudgery in farm operation with heavy reliance on traditional farm implements and methods of production contributing to low productivity.

vi) Low level of private sector investment in large scale agric production including low capacity for processing agric products to add value prior to selling, which is made worse by low producer prices and low access to credit facility.

vii) Poor state of existing grazing reserves and Low genetic potentialities of our existing livestock breed.

2.2   Empirical Literature

Sunusi et *al.,* (2006), Measures households' food insecurity in Lagos and Ibadan Metropolis in Nigeria. The study employs the use of descriptive statistics where frequencies, means, and standard deviation in analyzing data. The researchers observed that the major determinants that affect food security in the study areas are: income of the households, education level of the households and household size. The study also revealed that 26% of the household are food secure which are mostly within the teachers while 70% are food insecure.

Ojogho, (2007), examines the Determinants of Food Insecurity among arable farmers in Edo State of Nigeria. The researchers' employs the use of Multiple Binomial Logit Regression Model in analyzing the data collected. The model employed by the researcher is fitted with eight variables; household sex, household education, household size, household land holdings, household dependency ratio, age, per capita income of the household and total land cultivated.

However, the findings revealed that the probability of household being food insecure in the study area was 99.7% hence, approximately, three in every four people were found to be food insecure. The researcher identified that education level of farmers, household size, output level of households, and per capita income of households as the major determinants of food Insecurity in the study area. The paper recommended that a high level of educational level is needed to ensure food security status of households in the study area.

Agba, *et-al* (2009): conducts a research on Poverty, Food Insecurity and Rebranding Question in Nigeria. The paper dwells on descriptive analysis where it takes a critical looks at the current projects in the light of mounting predominance of poverty and food insecurity. The finding of the analysis recommends that rebranding should begin by addressing the leadership question, poverty, food insecurity, corruption and the decay in Nigerian educational system.

41

Martins, *et-al* (2009): Undertake a research on The State of politics of Poverty and Food Insecurity in Nigeria. The study uses descriptive analysis in explaining food insecurity crisis in Nigeria. Based on the analysis the study recommends the evolution of democratic power from civil society to pave way to policies which do not excludes the power. This should place the poor directly in charge of poverty alleviation programs, build agricultural capacity and subsequently increase food production.

Shifewa, (2003); conducted a research on the Determinants of Food Insecurity in Southern Ethiopia where the paper uses descriptive Statistics and Logit regression model to identify the determinants of food security in the study area. However, the findings revealed that seven out of nine variables used are statistically significant determinant of food security in the study area.

Okwudilio, *et-al* (2006): Conducted a research on Analysis of the Determinants of Food Insecurity with severe hunger in selected southern States. The study employs the use of two stage process involving the application of Rasch Measurement and Logit Model in analyzing the data. The study shows that both household with children and those without children income is the significant determinant of food insecurity in the study area.

Fantaw, (2007): Undertook a research on Food Insecurity and its Determinants in Rural Household of Amhara Regional State of Ethiopia. The study employs both descriptive and Econometrics Analysis, specifically Logit Regression Analysis. The findings of the study reveal that 45% of the households in the study area were found to be food Insecured. The stud y further reveals that household size, education, agric income, lives stock possession and off farm income are found to be statistically significant determinants of food insecurity in the study area.

Bogale, (2009): Undertook a research on Household level Determinants of Food Insecurity in Rural Areas of Dire, in Eastern Ethiopia. The study employs the use of binary choice techniques in identifying the factors influencing Household level Food Insecurity. The result obtained shows that family size, annual income, amount of credit received, access to irrigation, age of the households, farm size and livestock are theoretically consistence and statistically significance in explaining food insecurity situation in the study area whereas, sex of the households, off farm income, and education of households were found insignificant in determining food insecurity of the households in the study area. In addition, the paper recommends that food security situation need to build assets, improve the functioning of financial market and promotes family planning.

Maria, (2009): Conducts a research on a Country Perspective on the Determinants of Food Insecurity in Sub-Saharan Africa. The paper employs the use of Ordinary Least Square Method using cross sectional data of forty sampled sub-Saharan Africa in investigating the determinants of food insecurity at the macro level. The result of the findings provide useful information for rethinking food security policies and programs beyond the prevailing perspective of food availability and regional approach to the food crisis and considering the several dimensions of the concept and its differences between countries.

Makombe *et al.*,(2010); Undertake a research on Determinants of Food Insecurity in Malawi. The study employs descriptive statistics in identifying the determinants of food insecurity in rural Malawi. However, the study identified key policies that can reduce food insecurity: expanding the use of modern inputs to increase agricultural land productivity; increasing investment in road infrastructure to improve market access; expanding irrigation, agricultural extension activities, and social safety net programs; and investing in skills training and education for farmers. Policies aimed at reducing cost of food and farm inputs were also shown to reduce the probability of food insecurity.

However, the major findings revealed that the three determinants of food security are food availability, food accessibility and food absorption. Food availability includes the production of wheat, rice, maize, pulses, oilseeds, milk and meat. The components of food accessibility revealed that electrification and adult literacy rate contributes positively towards food security in the area and for food absorption the result shows that child immunization, female literacy, safe drinking water and hospitals positively affect food absorption in the study area.

In summary, almost all the reviewed work employed the use Probability Models in analyzing the possible determinants of food insecurity in their respective study areas. However, some of the researchers employed the use of multiple regression analysis applying Ordinary Least Square Method (OLS) and others uses less frequent methodologies/ estimation techniques like Linear Programming in presenting the food security/insecurity situation in their study areas.

Therefore, the above conclusion made by the above researchers' is reached due the various methodology employed and the variables used in the model. However, with inclusion of other different variables in the model, the results may likely change.

2.3 Theoretical Frame Work

It is based on the work Thomas Robert Malthus(1776-1834): A British Economist and Demographer, as well as political economist whose famous Theory of Population, highlighted the potential dangers of overpopulation in his famous work on An Essay on the Principles of Population, where he argued that population had a Natural growth rate describe by Geometric Progression i.e. 2,4,8,16. On the other hand the Natural resources necessary to support the population grew at a rate similar to Arithmetic Progression i.e. 1, 2,3,4,5 etc.

However, Malthus further concludes that without moral restraints, therefore, there will be continued pressure on living standard, both in terms of input and output. He was more particular to agricultural product because of the fear of hunger and famine due to the land fixation as compared with the size of population. According to him the transition of the two growth rates would mean that, population would soon outstrip the available resources and signified disaster. He therefore, prophesized of checks (natural and moral) will control population like hunger, famine, war, abstinence from sex and so on.

In line of the above Malthusian theory, Jigawa State is the 8th most populous State in Nigeria, with its rapid growing population rate of about 4.3 million people (Census, 2006) but the state is one of the poorest in the country with over 90.9% of the it citizens afflicted by poverty (NBS, 2007). However, poverty is the main cause of hunger and malnutrition which are aggravated by rapid population growth, hence poor are known to have inadequate consumption and they are limited to growth and brain development (Nord, 1999).

In addition, Malthus failed to recognize the presence of modern science and technology as well as international trade, but despite the low calorie intake, higher food prices, important food indices manifested in Nigeria and Jigawa in particular and hence violated the theoretical preposition of Malthus, in reducing food

45

insecurity especially in the immediate past. Beside, even where the country in question is densely populated like China and India, (even though china has taken a population control measure of one child per family) they still produce enough and even export the remaining surplus to other countries in the world.

In view of the above Malthus postulates that food insecurity if the function of Natural resources, Population growth and Moral restraints. In addition, Natural Resources and Moral Restraints exhibit negative relationship where as population Growth exhibit positive relationship with the level of food insecurity respectively.

Despite the shortcomings of the theory, the theory still serves as the theoretical base of this research study.

CHAPTER THREE
Research Methodology
3.1   Introduction
This chapter Contains the area of the study, methods of Data collection, population/sample size of the study, estimation techniques which is splited into; descriptive and econometric methods, followed by model specification, and data sources and measurement as well as definition of the variables.

3.2   Study Area
Jigawa State is situated in the North-western geo political part of the country between Latitudes $11.00°$ N to $13.00°$ N and Longitudes $8.00°$ E to $10.15°$ E. Kano and Katsina States border Jigawa to the west, Bauchi State to the east and Yobe State to the northeast. To the north in Maigatari Local Government - the home of the State's Export Free Zone Project, Jigawa State shares an international border with the Republic of Niger. This provides unique opportunities for international trade. (JSEEDS II, 2009)

Jigawa state, with a population of about 4.3 million people (Census, 2006), is blessed with large expanse of agricultural land, rivers and flood plains suitable for crops, livestock and fish production. Out of the 2.24 million hectares total land area of the state, about 1.6 million hectares is cultivable during the rain-fed season while about 308,000 hectares of the land mass is cultivable during the dry season through irrigation. Over 90% of Jigawa state adults solely depend on agriculture as a means of livelihood (CBN, 1999, cited in Abubakar & Ahmad, 2010).

The State has a total landmass of 24,742 square kilometres. A large proportion of this is certified to be arable. Ground survey data from JARDA, indicated that Jigawa State has a total Fadama (wetlands) land size of 3,433.79 square kilometres (one of the highest in the country). The land in the north-eastern fringes of the state, particularly Birniwa, Maigatari and Babura Local Government Areas, have the characteristics of the arid desert and is under threat of desertification.

47

Generally, the topography is characterized by undulating land, with sand dunes of various sizes spanning several kilometres in parts of the State. The southern part of Jigawa comprises the Basement Complex while the northeast is made up of sedimentary rocks of the Chad Formation (SEEDS II, 2007).

3.3   Method of Data Collection

In this study, primary sources of data are employed. The primary data were collected using the mixture of structured and unstructured questionnaire in order to capture the food insecurity of households in the study area.

3.4   Population/Sample Size of the Study

The study covered six local government areas of Jigawa State of Nigeria. However, a multi stage sampling was employed in selecting respondents for the study. Firstly, two local governments were selected from each Senatorial District, namely:

(1) North East; Auyo, *Birniwa*, Guri, *Hadejia*, Kafin- Hausa, Kaugama, Kiri-Kasamma, and Malam Madori.

(2) North West; Babura, Gagarawa, Garki, *Gumel*, Gwiwa, Kazaure, Maigatari, *Ringim*, Roni, Sule-Tankar-Kar, Taura, and ʻYan-Kwashi.

(3) South West; *Birnin-Kudu*, Buji, *Dutse*, Gwaram, Jahun, Kiyawa, and Miga, making six local governments out of twenty seven local government in the state.

Secondly, two communities each were randomly selected from the selected local governments, and these are Birniwa Town and Birniwa Tasha, Dubuntu and Matsaro, Lautai and Nakota, Nassarawa and Sintillmawa, Kofar fada and Cikin Gari, and Yadi and Fagoji respectively. Lastly, one hundred households were also selected from each community making six hundred households as the sample of the study.

In addition, six hundred questionnaires were administered between Novembers - December, 2010 to the six selected local government of the study area. However, five hundred and fifty questionnaires were retrieved making the sample of the study.

3.5   Estimation Techniques:

The study used a combination of both quantitative and descriptive techniques in analysing the collected responses.

3.5.1 Descriptive Method

Descriptive statistics was employed to explain the social and demographic characteristic of the households in the study area. It was also be used to assess the level and extent of food insecurity problem in the Jigawa State. The specific methods of data analysis involved the use of simple tables and percentages.

3.5.2 Inferential Method

The study employed the use of Multiple Binary Choice Modelling Techniques (Specifically Logit and Probit) Regression analysis. Logit and Probit models are log-linear models that allow the mixture of categorical and continues variable with respect to continues dependent variable. The model employs cumulative distribution function where response variables are dichotomous taking   0 – 1 value. However, Logit and Probit Models were employed due to their striking similarities and the difference between them is that Logit uses logistic Cumulative Distribution Function (LCDF) while the Probit uses the normal Cumulative Distribution Function (NCDF.)

The conventional R^2 for measuring the goodness of fit is of secondary importance in binary regression models, the measure similar to this is the pseudo R^2, and there are a variety of them. In most studies the Mc Fadden R^2 donated by $R^2{}_{McF}$ is used and its values ranges from 0 – 1. Another comparatively measure of goodness of fit is the count R^2 which is defined as number of corrected predictions divided by the total number of observations. Goodness of fit in binary choice models are of secondary importance. What matters are the expected signs of the regression coefficients and their statistical/practical significance. Logit techniques is used in this study due to its relative simplicity as Guajarati (2007) observed that in practice many researchers choose the Logit model because of its comparative mathematical simplicity. While the question of which model to use in binary choice analysis is unresolved, it has been observed that in most applications, it does not make much difference since the models give similar results (Gujarati 2002).

3.6   Model Specification

The study adopts the model employed by Ojogo (2007), in identifying the Determinants of Food Insecurity among arable farmers in Edo State of Nigeria. The researcher adopted the Malthusian Theoretical model with some modifications. Thus; the model includes five more variables in addition to the three exogenous models as specified originally by Malthus.

However, in this study cost-of-calorie of food consumed by each household in Naira as a minimum calorie requirement will be use as a benchmark to differentiate between food secure household and food insecure household. This method is been used by many studies whose main focus is on food insecurity (Greer & Thorbecke, 1986; and Hassan & Babu, 1991).

In line with the above, a household whose daily estimated expenditure of food is less than ($1/₦157.00) per day is be considered as food insecure household, otherwise food secure household as specified by (FAO, 1996 & World Bank, 1996). To achieve this, household estimated total expenditure on food consume per day is divided by his family size to get per capita daily expenditure of a household.
In this case the dependent variable is food insecurity, a binary variable which took the value one if a household is found to be food insecure, zero otherwise. The Logistic Cumulative Distribution Function Model can be specified as:

50

$$Pi = F(Z_i) = \cfrac{1}{1+e^{-(\alpha+\Sigma\beta_i X_i)}} \qquad - \quad - \quad - \quad - \quad - \quad - \quad - \qquad 1$$

Where;

Pi= Probability of an individual being Food Secure

Xi= Represents the ith explanatory variable

α & βi are regression parameters to be estimated

e is the base of natural logarithm

Logit and Probit models are usually estimated based on the method of maximum likelihood. For ease of interpretation of the coefficients, a logistic model could be written in term of odds and log of odds. The odd ratio is the ratio of the probability that an individual or household that would be food insecure (*Pi*) to the probability of a house hold that would be food secure (*1-Pi*).

Suppose we have a random sample of n observations. Letting $f_i(Y_i)$ denote the probability that $Y_i = 1$ or 0 the joint probability of observing the n Y values i.e $f(Y_1, Y_2,.....,Y_n)$ is given as;

$$f(Y_1, Y_2,......,Y_n) = \prod_1^n f_i(Y_i) = \prod_1^n P_i^{Y_i}(1-P_i)^{1-Yi} \qquad\qquad 2$$

The equation is known as the log likelihood function (LLF). In maximum likelihood the objective is to maximize the LLF that is to obtain the values of unknown parameters in such a manner that the probability of observing the given Y's is high (maximum) as possible.

The parameters of the model, α and β_1, can be estimated using the maximum likelihood (ML) method.

However, this analysis will be more concerned with the modified version of the model and is defined as follows:

$$Z = \alpha + {}_1X_1 + {}_2X_2 - {}_3X_3 - {}_4X_4 + {}_5X_5 - {}_6X_6 + {}_7X_7 + \mu_i \qquad\qquad 3$$

Where: Z = Food Insecurity (1= Food Insecure & 0, Otherwise)

α = Constant

$_1$- $_7$ = Regression Coefficients.

X_1 = House Hold Income.

X_2 = House Hold Family Size.

X_3 = House Hold Farm Size (acres).

X_4 = Sex (1=Male & 0=Female).

X_5 = House Hold Education Status.

X_6 = House Hold Age.

X_7 = House Holds Engage in Farming (1=Farming, 0=otherwise).

μ_i = Error Term

3.7   Data Sources and Measurements

Household income (X_1): This refers to the monthly earnings of the households from his working environment. The income is expected to reduce households' food insecurity. The expected effect of this variable on food insecurity is negative.

Household Family size (X_2): The number of adult individual members in the household measures household size. Since food requirements increase with the number of persons in a household, the expected effect is positive.

Households Farm size (X_3): Farm size is the total farmland holdings of the household measured in acres. The larger the farm sizes the lower the insecurity. It is thus expected that households with larger farm size are more likely to be food secure than those with smaller farm size. The expected effect on food insecurity is negative.

Household head/ sex (X_4): The sex of the households determines whether they are male or female household. Thus, male household is less likely to be food insecure than female household. The expected effect on food insecurity for male household is negative and for female household is positive.

Households Level of Education (X_5): Educational attainment of individual households' determines their food security status. The expected effect of education/year of schooling of a household on food insecurity is negative. This is because the higher the education of a household, the lower the food insecurity and vice-versa.

Households Age (X_6): The age of household's head in year is expected to have positive impact on food insecurity. This is because the more the household is becoming older the more will be his food insecurity since he cannot work and earn income but rather depends on his past savings.

Household Engagement in Farming (X_7): This refers to the households that are engage in farming practices. The expected effect of this variable on food insecurity is negative. This is so because the more the households are engage in farming practices the lower will be the insecurity and vice-versa.

Table 1: Definition of the variables used in the estimated equation.

Variables	Definitions	Symbols used	A Priori Expectations
Household Food Insecurity (Dependant)	1 = If Food Insecured 0= Otherwise	HFINS	
Explanatory Variables			
Households' Income (X_1)	Households' Monthly Income	HINC	$X_1 < 0$
Households' Family Size (X_2)	Number of Dependants per Households'	HFSZ	$X_2 > 0$
Households' Farm Size (X_3)	Size of Households' Farm Holdings in Acres	HFMS	$X_3 < 0$
Households' Sex (X_4)	1 = If Male 0= Otherwise	HSEX	$X_4 < 0$
Education of Household/ Years of Schooling (X_5)	0= Informal Education 6= Primary Education 12=Secondary Education 16=OND/NCE 18=BSC/HND 20 + =Post Graduate Studies	HEDU	$X_5 < 0$
Household Age (X_6)		HAGE	$X_6 > 0$
Household Engage in Farming (X_7)	1= If Engage in Farming 0= Otherwise	HEFM	$X_7 < 0$

Source: Researcher's Extraction, 2010.

54

CHAPTER FOUR
Data presentation and analysis

4.1 Introduction

This chapter contains the presentation and analyses of the data collected from the respondents within the six local governments' areas of Jigawa state, Nigeria.

First part of this chapter deals with the presentation and analysis of responses using descriptive statistics to explain the socio-economic and demographic characteristics of the households while second section of the chapter deals with the inferential analysis using binary choice modeling techniques.

4.2.1 Table 1: Households' Sex, Educational Qualification and their Ages

Variables		Frequency	Percentages (%)
Household head/Sex of the household	Male	516	93.82
	Female	34	6.18
	Total	550	100.00
Households Level of Educational	Informal	15	2.73
	Primary	85	15.45
	Secondary	116	21.09
	OND/NCE	243	44.18
	BSC/HND	67	12.18
	PG Studies	24	4.36
	Total	550	100.00
Age of the House holds Heads	<30	149	27.09
	30 – 49	352	64.00
	50 – 69	36	6.55
	70 +	13	2.36
	Total	550	100.00

Source: Field Survey, 2010.

The above table shows that 516 out of 550 sampled households which constitute 93.82% were Male headed households, where as the female headed household stood at 34 out of 550 which account for 6.18%. The possible reason for the above dispersion perhaps could be due to culture, religion and tribe of the respondents hence, they are predominantly Hausa Fulani.

55

The same table revealed that only 15 respondents out of the 550 sampled households are said to acquire informal type of education where 85 and 116 of the respondent acquire primary and secondary school certificates which constitutes 15.45 and 21.09 percents respectively. Furthermore, 243 and 67 of the respondents acquires OND/NCE and BSC/HND certificates which constitute 44.18 and 12.18 percents respectively, while the remaining 24 respondents which constitute 4.36 percents acquire both masters' degrees and postgraduate Diploma certificates.

The basic reason for why NCE/OND holders are the majority in the study area could perhaps be due to the establishment of the state College of Education and the state Poly Techniques as the only higher institutions of learning in the study area then.

However, the same table further depict that, 149 respondents are below 30 years of ages which account for 27.09 percent. In addition 352 and 36 respondents are between the age of 30 – 49 and 50 – 69 years which account for 64.00 and 6.55 percents respectively. Similarly, 13 out of 550 respondents are within 70 years of age and above in the study area which constitutes 2.36 percent.

Given the above analysis, we can clearly deduce that majority of the respondents in the study area falls within the working class with very few of them employed in the informal sector of the economy.

4.2.2 Table 2: Household Heads Marital Status.

Variables	Frequency	Percentages (%)
Married	470	85.45
Single	44	8.0
Divorced	16	2.91
Widowed	20	3.64
Total	550	100.00

Source: Field Survey, 2010.

As indicated from the above table, the married household account for 85.45% of the sampled household. Accordingly, Bachelors, the divorced and widowed accounted for 8.0, 2.91 and 3.64 percents respectively.

The possible reasons of the above dispersion may be due to the respondent's tribe, culture and religion.

4.2.3 Table 3: Households Occupation and Monthly Income.

Variables		Frequency	Percentages (%)
Occupation	Farming	140	25.45
	Trading/Business	76	13.82
	Salary/Wage Earners	326	59.27
	Others	08	1.45
	Total	550	100.00
Monthly Income	<N10,000	16	2.91
	N10,000 – 39,000	222	40.36
	N40,000 – 69,000	84	15.27
	N70,000 – 99,000	64	11.64
	N100,000 +	77	14.00
	Others	87	15.82
	Total	550	100,00

Source: Field Survey, 2010.

The above table portrays the various occupations and Monthly income of the sample household in the study area. Most of the respondents in the study area were Salary/Wage earners. This is because 326 out of 550 respondents were civil servants which accounts for 59.27% of the total responses. This is closely followed by Farmers, Traders/ Business Men and others who constitute 25.45%, 13.82% and

1.45% respectively. Furthermore, most of the households in the study area practice farming as their secondary occupation.

Besides, the Maximum earning capacity of the households in the study area stood between N100, 000+, N70, 000 – N99, 000 and N40, 000 – N69, 000 which accounts for 14.00, 11.64 and 14.36 percents respectively. Whereas the minimum earning capacity of the households stood between   N10, 000 – N39, 000 which constitute 40.36 percents respectively, thus, 87 household in the study area which account for 15.82 percent has no income.

4.2.4 Table 4: Household Estimated Farm Size in Acres.

Farm Size(acres)	Frequency	Percentages (%)
1.5	199	36.18
2	87	15.82
2.5	86	15.64
3	45	8.18
3.5	43	7.82
4 +	19	3.45
Others	71	12.91
Total	550	100.00

Source: Field Survey, 2010.

It is evident from the above table that; majority (203) of the sampled household possesses 1.5 acres of farmland which accounts for 36.18 percent out of 550 samples. However, 87, 86 and 45 respondents possess 2, 2.5 and 3 acres of farmland which accounts for 15.82, 15.64 and 8.18 percents respectively. In addition, 43 respondents out of the total sampled households have 3.5 acres of farmland which constitute 7.82 percent, whereas, 71 households which accounts for 12.91 percent possesses 4 and above acres of farmland in the study area, while others without farmland stood at 71 which account for 12.91 percent.

4.2.5 Table 5: Households' perception on the Causes of Food Insecurity.

Items	Frequency	Percentages (%)
Family size	124	22.55
Illiteracy	66	12.0
Poverty	336	61.09
Others	24	4.36
Total	550	100.00

Source: field survey, 2010.

The above table specified that 124 respondents which constitute 22.55 percent believed that family size causes food insecurity most in the study area, where as 66 out of 550 which account for 12.0 percent believed that food insecurity is caused most by illiteracy, while 336 respondents account for 61.09 percent believed that food insecurity is caused mostly by poverty and 24 of the sampled household believed that there are other reasons apart from the one mentioned above.

However, we can clearly deduce from the above analysis that apart from the specified determinants of food insecurity in the model, poverty is also one of the major factors that hinders food security in the study area. This is because 61 percent of the respondents strongly believed that poverty is the major cause of food insecurity in the state, hence this is in line with the findings of (Kopir, 2007) that poverty is the major causes of Hunger and Malnutrition which are aggravated by rapid population growth hence poor are known to have inadequate consumption and they are limited to brain growth and development.

4.2.6 Table 6: Household Family Size

Family Size	Frequency	Percentages (%)
1 – 5	175	31.82
6 – 10	179	32.55
11 – 15	87	15.82
16 – 20	67	12.18
21 – 40	26	4.73
No Children	16	2.91
Total	550	100.00

Source: field survey, 2010.

The minimum and the maximum family size of the households were found to be 26 and 179 respondents. However, 4.73 percent of the sampled households' have family size between 21– 40. Accordingly, 31.82 and 32.55 percents of the respondents have the family size between1-5 and 6 – 10 respectively, where as 15.82 & 12.18 percents of the respondents have the family size between 11 – 15 and 16 – 20 respectively. Furthermore, 16 of the respondents have no children which account for 2.91 percent.

However, given the standard of World Bank as earlier stated in chapter three (3) of the study, the poverty line is drawn to be an income of one dollar ($1) per person per day.

Therefore, this research study considers all household members whose daily income is above one dollar ($1) which is equivalent to (N157.00) is considered to be food Secured household, otherwise, food Insecure household, hence the following conclusions can be drawn:

- ❖ N157/3 = N52.33 per person per meal.
- ❖ From the above table, the average number of family size per household = 9.
- ❖ Average meal per family = N52.33 X 9 = N471
- ❖ Average meal per family per day = N471 X 3 = N1,413.00

4.2.7 Table 7: Estimated Cost Spent on Food by Household per day in Naira ₦).

Estimated Cost	Breakfast	%	Lunch	%	Dinner	%
< 157	338	61.45	244	44.36	210	38.18
157-300	69	12.55	150	27.27	162	29.45
301-470	45	8.18	114	20.72	138	25.09
471+	98	17.82	42	7.64	40	7.27
Total	500	100	550	100	550	100

Source: Field survey, 2010.

The above table indicates that, 338, 244 and 210 out of 550 sampled households which accounts for 61.45, 44.36 and 38.18 spent less than N157.00 per day on breakfast, lunch and dinner respectively. In addition, 69, 150 and 162 respondents which constitute 12.55, 27.27 and 29.45 percents spent between N157 – N300 per day on breakfast, lunch and dinner in their respective order.

Given the average family size per household to be nine (9) though they spent between N157 – N300, but when divided by the average number per household they are food insecure.

However, 45, 114 and 138 households which accounts for 8.18, 20.72 and 25.09 percent spends between N301 – N470 naira per day on breakfast, lunch and dinner, whereas, 98, 42 and 40 households spends N471.00 naira and above on food eaten per day which account for 17.82, 7.64 and 7.27 respectively.

4.2.8 Table 8: Classification based on Food Secure and Food Insecure Household.

Classification	Frequency	Percentage (%)
Food Insecured Households	452	82.18
Relatively Food Secured Households	58	10.55
Fully Secured Household	40	7.27
Totals	550	100.00

Source: Field Survey, 2010.

The above table depicts that, only forty (40) households are fully food secured throughout the day which accounts for 7.27 percent of the total sampled household in the study area.

Additionally, 58 respondents out 550 sampled households which constitute 10.55 percent are relatively food secured, this is because each of them is secured either once or twice out of the three square meals per day. Thus, the whole 98 households are secured during breakfast, only 42 household are food secured during lunch while 40 households are fully secured throughout the day.

Furthermore, 452 households out of 550 sampled households which account for 82.18 percent in the study area are found to be food in secured households.

However, the above result conforms to the poverty study conducted by National Bureau for Statistic (2012), which shows that the state has 79.0% level of poverty.

4.3 Inferential Estimate of the Likelihood of Household Food Insecurity.

From the above analysis, it is possible to classify the households under the study area as food insecure households. This is because 452 households out of 550 sampled household (82.18%) are food insecure simply because they cannot afford to spend one dollar on one square meal per person per day. Thus, only 98(17.82) of the sampled household could afford to spend one dollar ($1) per meal per persons per day.

However, finding the factors that determines the likelihood of being food insecure or otherwise goes beyond descriptive analysis and requires employing econometrics analysis as it was earlier mentioned in methodology.

Based on specification of the model in methodology of the study, Logit model was estimated to see the effect of the expected determinant factors of food insecurity in the study area.

The results are presented in the table below. The magnitude of the coefficient obtained from the model show the effect of each explanatory variable on the probability of being food insecure or otherwise.

4.3.1 Logit Estimation Results of Food Insecurity Status of the Households.

Variables	Coefficient	Std. Error	Prob.
C	3.358324	0.909525	0.0002
$HINC(X_1)$	7.450006	3.070006	0.9806
$HFSZ\ (X_2)$	0.014256	0.019197	0.4577
$HFMS(X_3)$	-0.142075	0.111753	0.2036
$HSEX(X_4)$	-0.809833	0.622452	0.1932
$HEDU\ (X_5)$	-0.081234	0.029904	0.0066
$HAGE(X_6)$	0.018014	0.010924	0.0991
$HEFM(X_7)$	-0.609177	0.271044	0.0246
Count $R^2 = 0.22$	H-L= 4.6549	Andrew=5.1950	

Source: Researcher's Computation using Eviews (3.0) Software.

63

$$Z = 3.358324 + 7.45008_1X_1 + 0.014256X2 - 0.142075X_3 - 0.809833X_4 - 0.81234X_5 + 0.018014X_6 - 0.609177X_7$$

The above table shows the estimation of the determinants of food insecurity as it was specified in the model. However, the coefficient of the constant term stood at 3.358324, which implies that holding other variables constant food insecurity in the study area stood at 336%.

In addition, the coefficient of Household Income (*HINC*) in the study area as one of the determinants of food insecurity was found to be statistically insignificant and did not conforms to apriori expectations. This is because *HINC* coefficient stood at 7.4500 and this implies that a unit increase in Households Income on average tends to increase the probability of Households Food Insecurity by 745%, hence the coefficient of income carries a positive (+) sign indicating that the higher the Households Income, the higher the Insecurity and vice-versa.

The possible reason for the above positive coefficients perhaps could be due to the fact that most of the household with higher incomes in the study area do not attach much interest in buying food that will balance their diet but rather attach more interest on buying assets like landed properties, cars and getting married frequently as their income increases, hence majority of the respondents in the study area earns lower income which can also make the variable to be positive indicating low consumption on food items.

In terms of odds, the antilog of the coefficient of Household Income ($e^{7.450006}$) $\approx$ 1719.87, this means that households with higher incomes are 1719.87 times less likely to be food insecure then households with lower incomes, other things remain constant.

However, in line with researcher's expectation, the Household Family Size (*HFSZ*) has a positive coefficient of *0.0142075*. This implies that, a unit increase in

Households Family Size (Dependents) on average tends to increase the probability of households food insecurity by 1.42%, hence, the higher the household's family size the higher level of food insecurity and vice-versa. In addition majority of the households in the study area have family size of 1-5 and 6-10 respectively, with nine members on average per household.

In terms of odds, the antilog of the coefficient of Household Family Size ($e^{0.014256}$) $\approx$ 1.01, this means that households with larger family size are 1.01 times more likely to be food insecure then those household with lower family size.

The third variable in the regression coefficient was the household farm size (*HFMS*). The above result shows a negative coefficient of -0.142075. This implies that a unit increase in households' farm size on average tends to decrease the probability of households food insecurity by 14.21%. In addition this variable conforms to apriori expectation. This is because the higher the farm size, the higher the quantity of food grown and the lower the household food insecurity and vice-versa. Even though majority of the respondents in the study area possess land holdings of 1.5 acres, thus among them some are cultivating it while some are not.

In terms of odds, the antilog of the coefficient of Households Farm Size ($e^{-0.142075}$) $\approx$ 0.87, this means that households that acquire farmland are 0.87 less likely to be food insecure than those households that did not possess farmland.

In addition, it is assumed that in most societies, the head of a household strongly influences the household's livelihood and food security. Thus, the household Sex (*HSEX*) as one of the determinants of food insecurity in the study area as specified in the model conforms to apriori expectation. This is because the variable has negative sign though statistically insignificant with the coefficient of *-0.809833*, which implies that male headed households have higher probability to be less food insecure than their female counterparts, hence the majority of the household heads in the study area were found to be male headed households.

In terms of odds, the antilog of the coefficient of Households Sex ($e^{-0.809833}$) $\approx$ 0.44, this means that male headed households are 0.44 less likely to be food insecure than their female counterpart.

On the other hand, the variable of household educational attainment/years of schooling variable has a negative significant coefficient of *-0.081234*. This implies that a unit increase in household educational attainment/ years of schooling on average tends to decrease the probability of households food insecurity by 8.12%. This is because the respondents with higher educational attainment/ years of schooling are the ones with higher incomes in the study area. However, variable conforms to apriori expectation because the higher the education the lower the households food insecurity and vice-versa.

In terms of odds, the antilog of the coefficient of Household Education ($e^{-0.081234}$) $\approx$ 0.92, this means that the household with higher educational attainment / years of schooling are 0.92 less likely to be food insecure than those households with lower educational attainment/years of schooling.

From the above table, the result with regards to Age (*HAGE*) of the household in the study area has a positive coefficient of *0.018014*. This implies that a unit increase in the household's age on average tends to increase the probability of household food insecurity by 1.8%.

The possible reason could be as the household heads becomes older, their chances to work and earn income becomes difficult and hence, depends on past savings or on their children and relatives.

In terms of odds, the antilog of the coefficient of Household Age ($e^{0.018014}$) $\approx$ 1.02, this means that households with higher age are 1.02 more likely to be food insecure than those households with lower age.

66

The last variable of household food insecurity as specified in the model of the study is households that are engage in farming practice (*HEFM*). This variable conforms to apriori expectation because it has a negative but insignificant coefficient of -0.6091. The possible interpretation derive from this could be a unit increase in households that are engage in farming practices on average, tends to decrease probability of households food insecurity by 60.91%.

In terms of odds, the antilog of the coefficient of Household that are engage in Farming ($e^{-0.609177}$) $\approx$ 0.54,, this means that households that are engage in farming practice are 0.54 less likely to be food insecure than those households that did not engage themselves in farming.

In addition, all the regressors have a significant impact on food insecurity, hence the likelihood ratio (LR) statistics is 20.12537 with *P* value of 0.005305 (see appendix III).

The count R^2 gives the actual and predicted values of the regressand. From the fitted table in appendix III we can observe that out of 550 observations there were only 98 incorrect predictions. In addition, the count R^2 98/550=0.18 suggesting that 18% of the variability is explain by the model, where as McFadden R^2 value from the estimated regression model in appendix III is (0.0390). Although the two values are not directly the same but only gives idea about the orders of magnitude (Gujirati, 2007).

However, a goodness of the fit test was conducted using Hosmer-Lemeshow & Andrew Statistics. The decision rule for H-L & Andrew test says if H-L & Andrew is > 0.5, then it exhibit a good fit, otherwise bad fit. The result of H.L is (4.65) with probability of 0.7937 where as Andrew Stood at (5.20) with probability of 0.8778. This shows that a regression exhibit a good fit hence both H-L & Andrew are greater than 0.5. Eviews 3.0 help topic.

CHAPTER FIVE
Summary, Conclusion and Recommendation

5.1 Summary

Food insecurity is said to exist when people at all times do not have Physical and Economic access to adequate, safe and nutritious food to meet their dietary requirement and food preferences for an active and healthy life. Food insecurity problem in Nigeria can be traced back during the colonial era where colonial masters introduced the concept of taxation making our people to divert from subsistence farming to cash crop production in order to sold the produce and pay taxes. Consequently, neglect of Agricultural sector due the discovery of oil worsens the food insecurity problem in Nigeria up till today where most of the populace cannot afford three square meals. It has been observe that a lot of studies of this nature were carried out in Nigeria and outside Nigeria, but unfortunately no any research of this kind was conducted in the study area, thus, it is in light of the above, this research was carried out.

The problem of food insecurity is pervasive in the study area. The study aimed at identifying the determinants of food insecurity and identifies possible ways of tackling such problems.

Accordingly, the study employed primary sources of data where the data collected was analyzed through descriptive statistics and binary choice model techniques.

The result obtained from the analyzed data revealed that among the 550 sampled households, 452 respondents which account for 82.18% in the study area were found to be food insecure households while 98 out of the total sampled households which constitutes 17.82% are found to be food secured households out of which 58(10.55%) and 40(7.27%) households are relatively and fully food secured respectively. Thus, only 40(7.27%) of the sampled households could afford to spent

68

($1 = N157.00) or more per person per day on three square meals i.e. Breakfast, lunch and dinner.

However, the percentages of those with OND/NCE (44.18%) are higher in the study area followed by secondary (21.09%), and primary (15.45%) schools respectively. Those with BSC/HND, Post graduate certificates and those with informal type of education account for 12.18, 4.36 and 2.73 percents respectively.

Furthermore, 336(61.09%) of the respondents believe that food insecurity is caused most by poverty, while 124(22.55%) believed that food insecurity is caused most by the family size of the household and 66(12%) believed with illiteracy as the major cause of food insecurity in the study area. In addition, male headed households dominate the sample size of the study with 93.82 percent whereas female headed households account for only 6.18 percent.

Majority of the households in the study area are civil servants (326), followed by (140) farmers, (76) trader/business men while (8) of the respondents are not working in the study area. Most of the results obtained in descriptive analysis are consistent with the results obtained from the regressed model. The model was fitted with seven (7) explanatory variables *(HINC, HFSZ, HFMS, HSEX, HEDU, and HEFM)*. However, three out of seven variables i.e. *HEDU, HAGE, and HFMS*, were found to be statistically significant carrying the expected signs, and conforms to apriori expectations.

In addition, four out of seven variables used in the model (*HINC, HFSZ, HFMS, and HSEX*) were found to be statistically insignificant but conforms to apriori expectation hence they are carrying the expected signs as specified in the model with exception of households income. Perhaps, the possible reason that makes household income to carry unexpected sign could be due to the fact that most of the households in the study area do not attach more value to buy food stuffs that will balanced their diet but rather attached more value in buying more landed properties, cars and getting married.

In light of the evidences obtained from the regression results, household family size*(HFSZ)*, household farm size *(HFMS)*, household sex *(HSEX)*, household education *(HED)*, household age *(HAGE)*, and household engagement in farming *(HEFM)*,conforms to apriori expectations, thus; only household income *(HINC)* did not conform to the apriori expectation as specified in the model.

In view of the above, household education *(HEDU)*, age of the household *(HAGE)* and household engagement in farming *(HEFM)* were found to be statistically significant determinant of food insecurity in the study area at 1%, 10% and 5% respectively, where as households income *(HINC)*, household family size *(HFMS)*, household farm size *(HFMS)*, and household sex *(HSEX)*, were found to be statistically insignificant determinants of food insecurity in the study area,

In addition, the result obtained from the analyzed data (Descriptive & Inferential) revealed that food insecurity is pervasive in the study area. This is because among the 550 sampled households, 452 (82.18%) of the respondents in the study area were found to be food Insecured households while only 98(17.82%) of the respondents were found to be food secured households.

However, this research study is consistence with the poverty assessment conducted by National bureau for Statistics that 79.0% of the study population was afflicted by poverty (NBS, 2012).

5.2 Conclusion

Based on the evidences obtained from the regression results, the study concludes that three variables household education *(HEDU)*, household age *(HAGE)*, and household engagement in farming *(HEFM)* out of the seven variables fitted in the model were found to be the major determinant of food insecurity in the study area this is because they are found to be statistically significant determinant of food insecurity in the study area.

In addition, 452 (82.18%) and 98 (17.83%) were found to be food Insecured and food secured households respectively. This leads to the conclusion that 82.18% of the households in the study area were found to be food Insecured.

In view of the above, the study conforms to Ojogo's (2007) study in one way; household education was found to be statistical significant determinant of food Insecurity in the study area.

However, the study varies with Ojogo's findings in two ways; household age *(HAGE)* and household engagement in farming (*HEFM*) are another significant determinant of food insecurity in this study where as Ojogo (2007); household size, output level of household and household per capita income are identified as the major determinant of food insecurity instead. The above variation perhaps could be due to the characteristics of the study areas.

5.3   Recommendations

The followings are the possible recommendations based on the findings of the research study;

From the above regression results households education *(HEDU* is one of the major determinants of food insecurity in the study area, therefore, government should provide sound and conducive atmosphere of teaching learning processes by investing massively on education in order to promote the households education thereby, addressing food insecurity problems in the state, hence nowhere in the world an educated person is really poor.

Households' engagement in farming *(HEFM)* is also a significant determinant of households' food insecurity in the study area. In view of the above, government should enhance agricultural programs that will encourage households irrespective of their occupation to make farming their secondary occupation making them to cultivate available farm land at their disposal, through the provision of fertilizers and higher yielding variety seeds *(HYV's)* by giving them appropriate subsidy and farm input were necessary in order to reduce the food insecurity problems in the study areas.

However, food insecurity according to the collected responses of the households in the study area is caused mostly by poverty. In view of the above, Government should design a policy that will directly empower the households in the study area in order to do away with food insecurity problem and reduce the magnitude of poverty in the study area as well.

More researches could be developed in this area to identify more determinants of food security in the study area. Other areas of further research include the impact of climate on Food Security.

BIBLIOGRAPHY

Abdulla Y.I (2007): An Assessment of the Causes of Food Insecurity in   Southern African Stellenbosch University. Unpublished thesis

Abubakar M.S & Ahmad D., (2010); Utilisation of and Constraints on Animal Traction in Jigawa State Australian Journal of Basic and Applied Science. INSInet Publications.

Adeyeye, V.A., (1997). Food and Nutrition Survey in Nigeria. 1st Eition. University of Ibadan Press. Ibadan, Nigeria.

Agba M. Micheal S. UJshei E. And Akwara F., (2009): Poverty Food Insecurity and Rebranding Question in Nigeria. Published by Canadian Academy of Oriental and accidental Culture. Vol 5 No. 6 http//www.cscanada.org.net

Aredo D. (1995). "Transforming Peasant Agriculture: Conceptual Framework". In Demeke M. and Aredo D. (editors). *Problems and Transformation*, proceedings of the Fourth Annual conference on the Ethiopian economy.

Bogale A. And Shimelis A. (2009): Household Level Determinants of Food Insecurity in Rural Areas of Dire- Diwa, Eastern Ethiopia. Ifand online Journal Vol. 9 No. 9

Campbell,C. (1991). Food insecurity: a nutritional outcome or predicator variable? Journal of Nutrition; 121; 408-415

Central Bank of Nigeria, (2005): Annual Report and Statement o Account CBN, Abuja, Nigeria.

Census, (2006): National population commission, Nigeria.

Central Bank of Nigeria, (2007); Statistical Bullion Vol. 31 N0. 4

Fantaw S.F., (2007), Determinants of Food Insecurity in Rural Households in Amhara Regions. Unpublished thesis Submitted to Adis Ababa University.
Food and Agriculture Organization of the United Nations, (1996); *World Food Summit*, corporate Documentary Repository, Rome, Italy.

Food and Agriculture Organization of the United Nations, (2001): *The State of Food Insecurity in the World.* Third edition. Rome: FAO, 2002.

Food and Agriculture Organization of the United Nations (2002). *The State of Food Security in the World,* 4th edition. Available from <http://www.fao.org>

Food and Agricultural Organisation (2005). The State of Food Insecurity in the World.

Greer, J. And Thorbecke, E. (1986): A methodology for measuring Poverty applied to Kenya. *Journal of Development Economics, 24 (I).*

Hassan, R. M. And Babu, S.C. (19991): Measurement and determinant of Rural Poverty; household consumption patterns and food poverty in Rural Sudan 16(6).

Idris, M. (2009): An Assessment of Household Food Insecurity in selected LGA of Kano State. A Ph.D Dissertation Thesis, in the Department of Economics, B.U.K

Igene J.O. (1997); Food production and Nutrition in Nigeria *Integrated Agricultural Production in Nigeria: Strategies and Mechanisms for Food Security.* Shaib, B., N.O. Adedipe, A. Aliyu and M.M. Jir (eds) The NARP and the FMANR.

Koutsoyiannis A., (2004): Theory of Econometrics. Second Edition Palgrave Publishers New York.

Kwanashie, M. (2010): First National Conference organized by the Department of Economics, Bayero University, Kano on "State of Nigerian Economy" at Musa Abdullahi Auditorium, New Site.

Maria S. (2009): A Country Perspective on the Determinants of Household Food Insecurity in Sub-Saharan Africa. Università degli Studi di Pavia

Martins I., Ngboawaje D. and Josephine N. (2009): The State of Politics of Poverty and Food Insecurity in Nigeria. International Bulletin of Business Administration, Euro Journals. http//www.eurojournals.com/IBBA.htm

Makombe T. Paul L., and Monica F., (2010); Determinants of Food Insecurity in Malawi. International Food Policy Research Institute. Sustainable Solution for Agriculture.

Mohammed I. (2008): Road map to attaining food security in Nigeria. Nigerian Tribune Newspaper, 25th March, 2008.

National Bureau for Statistics, (2012); Nigerian Poverty Profile Assessment Report.

NEEDS, (2004): National Abuja: National Planning Commission.

Nigeria, (2005); Nigerian Master Web http://www.nigeriamasterweb. com/

Nord, M., Jemison K. & Bickel, G. (1999), *Prevalence of food insecurity And Hunger, by state 1996-1998*, Food Assistance and nutrition Research Report No. 2 Food and Rural Economics Division, Economic Research Services, United State Department of Agriculture.

Nord, Mark; Andrews, Margaret; and Carlson, Steve (2002). *Household Food Security in the United States, 2001.* Food Assistance and Nutrition Research Report No. 29. Economic Research Service, U.S. Department of Agriculture. Available from <http://www.ers.usda.gov>

Okwuidilo O., Onoanwa C., and Gerald C., (2006): An Analysis of the Determinants of Food insecurity with severe hunger in selected Southern States. Published by Southern Rural Sociology.

Ojogho O. and Imodu P. (2007); Determinants of Food Insecurity Among Arable Farmers in Oredo and Egor Local Government Areas of Edo State, Nigeria: Policy Implication for Mitigating Food Crises.

Olayemi J.K. (1998); *Food Security in Nigeria*. Research Report No 2. Development Policy Centre, Ibadan.

Omotesho O.A, Adewumi M.O, Mohd A., and Ayinde O.E (2005); Determinants of Food Security Among the Rural Farming Households in Kwara State, Nigeria. African   Journal of General Agriculture Vol.2, No.1. Nigerian Publishers.

Oriola E. O. (2009): A Framework for Food Security and Poverty Reduction in Nigeria. *Department of Geography, University of Ilorin, Ilorin. Nigeria*

Orewa S. Iyangbe C., (2005): Household Food Insecurity in Nigeria: An Assessment of the Present status of Protein-Energy Malnutrition among Rural and Low-Income Urban Household in Edo State.

Osundare, F., (1999). Analysis of the Demand for Certified Maize Seeds in Ondo State, Implications for Food Security. A Paper Presented at the Nigerian National Association of Agricultural Scientist Conference.

SEEDS II, (2009); Jigawa State Comprehensive Development Frame Work Prepared by SEEDS II Technical Committee, Directorate of Budget and Economic Planning.

Shifewa F., (2003); Determinants of Food Security in southern Ethiopia: A Selected Paper Presented at American Agricultural Economics Association. Meetings in Montreal, Canada.

Sunusi R. Badejo C. and Yusuf B, (2006); Measuring Household Food Insecurity in Selected Local Government Areas of Lagos and Ibadan, Nigeria. Pakistan Journal of Nutrition.

USAID (1999).Food Security Indicators for the Use in the Monitoring and Evaluation of Food security Program.

U.S. Department of Agriculture (2002). *U.S. Action Plan on Food Security, Solutions to Hunger.* U.S. Department of Agriculture. Available from <http://www.fas.usda.gov>

World Health Organization. "Nutrition Research (2002): Pursuing Sustainable Solutions.. Available from <http://www.who.int/en>

World Bank (1986): Poverty and Hunger: Issues and Options for Food Security in Developing Countries. World Bank Policy Study. Washington D.C

World Bank Group. "Income Poverty, (2002): The Latest Global Numbers." Updated. Available from <http://www.worldbank.org>

Wood B, Wattanapenpaiboon N, Ross K, Kouris-Blazos A (2000): National Nutrition Survey: all persons 16 years of age and over – Food Security. Healthy Eating Healthy Living Program.

APPENDIX I

DATA USED IN THE RESEARCH STUDY

S/N	HFSS	HINC	HFSZ	HFMS	HSEX	HEDU	HAGE	HEFM	S/N	HFINS	HINC	HFSZ	HFMS	HSEX	HEDU	HAGE	HEFM
1	1	10000	1	0	1	12	27	1	50	1	0	11	1.5	1	18	50	1
2	1	10000	1	0	0	12	30	1	51	0	40000	6	1.5	1	18	50	1
3	1	10000	6	1.5	0	12	35	0	52	0	40000	6	1.5	1	18	50	0
4	1	10000	6	1.5	1	16	35	1	53	1	0	1	0	1	18	50	1
5	1	0	5	1.5	1	12	40	1	54	1	0	16	1.5	1	18	50	1
6	1	0	5	1.5	1	18	40	1	55	1	100000	21	1.5	1	12	50	1
7	1	0	5	1.5	1	18	40	0	56	1	40000	10	1.5	0	12	29	0
8	1	10000	15	1.5	1	12	30	1	57	1	40000	20	1.5	0	12	65	1
9	1	10000	5	0	1	12	30	1	58	1	69000	10	1.5	1	12	29	1
10	1	0	40	2	1	12	30	0	59	1	69000	5	3	1	0	29	1
11	1	0	10	0	0	12	30	1	60	1	0	10	3	1	6	27	0
12	1	10000	20	2	1	12	30	1	61	0	69000	21	3	1	6	27	0
13	1	0	5	2	1	20	45	0	62	0	0	6	2.5	1	6	23	0
14	1	10000	5	2	1	16	45	0	63	1	69000	1	2.5	1	6	27	0
15	1	39000	10	2	1	16	45	0	64	1	0	6	2.5	1	6	40	0
16	1	39000	40	2	1	16	45	1	65	1	0	6	3	1	6	40	1
17	0	39000	1	3	0	16	45	1	66	1	110000	6	3	1	6	40	1
18	1	39000	21	3	1	16	45	1	67	1	69000	6	0	1	6	40	1
19	1	39000	6	1.5	0	16	47	1	68	1	0	6	2.5	1	12	47	0
20	1	39000	1	0	1	20	48	1	69	0	69000	11	2.5	1	12	35	0
21	1	0	21	0	0	18	29	0	70	0	100000	5	2.5	1	12	35	0
22	1	39000	1	1.5	1	12	23	1	71	1	0	20	2.5	1	12	35	0
23	1	39000	6	1.5	1	12	25	1	72	1	10000	10	2	1	12	43	0
24	1	0	16	2.5	1	16	25	1	73	1	39000	21	1.5	1	0	39	0
25	1	39000	1	2.5	1	16	25	1	74	1	39000	5	0	1	12	37	0
26	0	70000	5	2.5	1	16	27	1	75	1	39000	10	0	1	16	37	0
27	1	70000	5	2.5	1	18	27	0	76	1	100000	10	1.5	1	16	37	1
28	1	70000	5	3	1	18	27	0	77	1	100000	5	1.5	1	16	37	1
29	1	70000	5	3	0	18	28	1	78	0	69000	10	1.5	1	16	37	1
30	1	70000	21	0	1	12	29	1	79	0	100000	21	1.5	1	16	43	0
31	1	70000	5	0	1	12	40	1	80	0	0	11	0	1	16	43	0
32	0	70000	16	2	1	12	40	1	81	0	100000	6	1.5	1	16	43	1
33	0	70000	1	2	1	16	40	1	82	0	0	6	1.5	1	16	43	1
34	0	70000	21	2	1	16	40	0	83	0	39000	1	3.5	1	16	43	1
35	1	70000	6	2	1	16	40	0	84	1	39000	11	3.5	1	16	44	1
36	1	70000	21	0	1	16	40	0	85	1	100000	6	1.5	1	16	44	0
37	1	70000	21	2.5	1	16	40	0	86	1	69000	6	1.5	1	6	44	0
38	1	70000	6	2.5	1	16	37	0	87	1	10000	6	1.5	1	16	27	1
39	1	0	5	2.5	1	16	34	0	88	1	10000	1	1.5	1	12	25	1
40	1	0	5	2.5	0	16	34	1	89	1	10000	21	1.5	1	12	25	1
41	1	10000	5	2.5	1	16	34	0	90	1	10000	16	1.5	1	12	25	1
42	1	10000	5	2.5	1	16	34	0	91	1	15000	1	1.5	1	16	25	1
43	1	10000	10	2	1	12	34	0	92	1	0	21	0	1	0	25	0
44	1	10000	10	2	1	12	34	1	93	1	10000	6	0	1	12	25	1
45	1	10000	5	2	1	12	35	0	94	1	10000	1	1.5	1	18	25	0
46	1	0	10	2	1	12	35	1	95	0	15000	21	2.5	1	18	28	0
47	1	10000	6	1.5	1	16	50	1	96	0	15000	6	0	1	12	29	1
48	1	10000	1	0	1	16	50	0	97	1	0	6	2.5	1	12	29	0
49	1	10000	21	1.5	1	16	50	0	98	1	10000	6	2	0	12	29	0

Continuation..........

99	1	10000	1	2	1	16	29	1
100	1	39000	16	2	1	16	29	1
101	1	39000	6	2	1	16	29	0
102	1	100000	0	2	1	16	35	0
103	1	0	16	2	1	16	35	0
104	1	10000	6	2	1	16	35	0
105	1	69000	16	2.5	1	16	35	1
106	1	100000	1	2.5	1	16	35	0
107	0	100000	6	2.5	1	16	35	1
108	0	39000	6	1.5	1	6	35	0
109	1	39000	1	1.5	1	16	37	0
110	1	10000	6	1.5	1	16	37	1
111	1	10000	11	1.5	1	16	37	1
112	1	10000	15	0	1	16	37	0
113	1	0	5	1.5	1	16	37	1
114	1	15000	5	1.5	1	12	39	0
115	1	0	7	1.5	1	12	39	0
116	1	69000	0	1.5	1	12	29	1
117	0	15000	10	0	1	0	27	1
118	0	10000	10	0	1	12	27	1
119	0	100000	10	3.5	1	12	25	1
120	0	69000	5	3.5	1	12	25	1
121	1	100000	15	1.5	1	12	25	0
122	1	100000	15	1.5	1	12	25	0
123	1	0	10	1.5	1	12	25	0
124	1	39000	10	1.5	1	12	25	1
125	1	110000	4	1.5	1	12	47	1
126	1	10000	17	1.5	1	12	49	1
127	1	10000	7	0	1	16	27	0
128	1	10000	5	3	1	16	27	1
129	1	39000	5	1.5	1	16	27	1
130	0	39000	5	1.5	1	16	27	1
131	0	110000	10	1.5	1	16	27	0
132	0	0	10	1.5	1	16	27	1
133	0	40000	0	1.5	1	16	35	1
134	0	39000	5	1.5	1	12	35	1
135	1	39000	15	1.5	1	12	37	1
136	1	40000	10	1.5	1	12	37	0
137	1	0	15	1.5	1	12	37	1
138	1	7000	20	1.5	1	12	39	1
139	1	7000	5	0	1	12	39	1
140	1	7000	5	0	1	16	39	1
141	1	7000	16	0	1	16	39	0
142	1	7000	10	0	1	16	39	0
143	1	7000	5	2.5	1	16	33	0
144	1	7000	5	2.5	1	16	33	1
145	1	7000	5	3.5	1	0	33	0
146	1	69000	5	0	1	16	33	1
147	1	39000	5	1.5	1	16	33	1
148	1	39000	5	1.5	1	16	33	1
149	1	0	5	1.5	1	16	33	0
150	1	40000	11	1.5	1	16	33	0
151	0	40000	6	1.5	1	16	40	0
152	0	39000	0	1.5	1	16	40	1
153	1	39000	16	1.5	1	18	40	0
154	1	39000	1	1.5	1	6	29	1
155	1	150000	1	3.5	1	18	29	1
156	1	150000	1	3.5	1	18	29	0
157	1	69000	1	3.5	1	18	35	0
158	1	0	1	3.5	1	18	35	0
159	1	15000	11	2.5	1	18	35	0
160	1	69000	6	0	1	18	35	1
161	1	39000	16	2.5	1	18	35	0
162	1	39000	1	2.5	1	18	35	1
163	0	15000	16	4	1	18	35	0
164	0	15000	10	3.5	1	18	37	0
165	1	69000	15	4	1	18	37	1
166	1	39000	5	1.5	1	18	37	1
167	1	39000	5	1.5	1	16	37	0
168	1	11000	1	1.5	1	16	37	1
169	1	10000	17	1.5	1	16	39	0
170	1	69000	1	1.5	1	16	39	0
171	0	0	3	1.5	1	16	29	1
172	0	10000	0	1.5	1	0	27	1
173	1	20000	2	1.5	1	16	27	1
174	1	20000	13	0	1	16	25	1
175	1	20000	7	3	1	16	25	1
176	1	20000	13	3	1	16	25	0
177	1	15000	9	2.5	1	16	25	0
178	1	65000	7	0	1	16	25	0
179	1	35000	3	2	1	16	25	1
180	1	110000	8	2	1	16	47	1
181	1	35000	8	2	1	16	49	1
182	1	65000	13	2	1	16	27	0
183	1	0	5	2	1	16	27	1
184	1	12000	3	2	1	16	27	1
185	1	12000	16	2	1	16	27	1
186	1	65000	6	4	1	16	27	0
187	1	35000	1	2.5	1	16	27	1
188	0	10000	6	2.5	1	16	35	1
189	0	45000	1	2.5	1	16	35	1
190	0	35000	0	2.5	1	16	37	1
191	0	120000	15	2.5	1	16	37	0
192	0	65000	10	3.5	1	16	37	1
193	1	99000	5	1.5	1	16	39	1
194	1	0	5	1.5	1	16	39	1
195	1	99000	5	1.5	1	16	39	1
196	1	39000	10	1.5	1	16	39	0
197	1	99000	10	0	1	6	39	0
198	1	65000	20	1.5	1	16	33	0
199	1	39000	10	0	1	12	33	1
200	0	99000	6	2.5	1	12	33	0
201	1	15000	1	2.5	1	12	33	1
202	1	65000	6	2.5	1	12	33	1
203	1	35000	6	2.5	1	12	33	1
204	1	65000	6	2.5	1	12	33	0
205	1	0	11	3.5	1	0	33	0
206	1	35000	5	1.5	1	12	40	0
207	0	35000	20	1.5	1	12	40	1
208	0	35000	10	1.5	1	12	40	0

Continuation..........

209	1	35000	5	4	1	12	31	1	263	0	35000	9	1.5	1	16	44	0
210	1	65000	19	3.5	1	12	28	0	264	0	35000	9	1.5	1	16	44	1
211	1	95000	11	3.5	1	12	27	0	265	0	10000	9	3.5	1	16	49	0
212	1	35000	6	1.5	1	16	44	0	266	0	40000	1	2.5	1	16	70	0
213	1	12000	0	1.5	1	20	44	1	267	0	70000	15	2.5	1	16	35	1
214	1	65000	16	0	1	20	46	1	268	1	10000	15	2	1	16	77	0
215	1	35000	6	1.5	1	20	49	1	269	1	39000	10	2	1	16	65	1
216	1	95000	6	1.5	1	20	27	1	270	1	10000	10	2	1	16	45	0
217	1	0	1	3	1	16	28	1	271	1	65000	10	2	1	16	45	1
218	0	65000	6	3	1	16	33	1	272	1	0	5	2	1	6	45	0
219	0	95000	6	3	1	16	33	0	273	1	0	5	3.5	1	16	29	0
220	1	0	11	2	1	16	37	0	274	0	0	5	2.5	1	16	29	1
221	1	5000	6	2	1	16	39	0	275	0	70000	15	0	1	16	29	0
222	1	5000	11	2	1	16	50	0	276	1	10000	15	1.5	1	16	29	0
223	1	5000	1	3	1	16	77	0	277	1	40000	5	1.5	1	16	29	0
224	1	5000	16	3	1	6	33	1	278	1	10000	15	1.5	1	16	43	0
225	1	5000	16	3	1	6	27	1	279	1	10000	10	1.5	1	16	43	1
226	0	5000	6	3	1	16	27	1	280	1	10000	20	1.5	1	16	47	0
227	0	99000	1	3	1	16	67	0	281	1	39000	10	1.5	1	0	45	0
228	0	65000	20	1.5	1	16	33	0	282	1	39000	20	1.5	1	16	45	1
229	0	35000	15	1.5	1	16	35	0	283	1	70000	5	3	1	16	45	0
230	0	35000	10	1.5	1	16	35	0	284	1	10000	0	3	1	18	30	0
231	0	39000	10	1.5	1	12	35	0	285	1	40000	10	0	1	18	30	0
232	1	100000	15	1.5	1	12	35	1	286	1	0	20	0	1	20	59	0
233	1	69000	5	3.5	1	12	27	1	287	1	10000	15	0	1	12	59	0
234	1	39000	15	0	1	12	35	1	288	1	10000	10	0	1	12	45	0
235	1	70000	0	1.5	1	12	35	0	289	1	40000	11	2	1	12	45	1
236	1	150000	20	1.5	1	12	35	0	290	1	7000	5	2	1	12	45	0
237	1	0	10	1.5	1	18	25	0	291	1	7000	5	2	1	12	27	0
238	1	35000	15	1.5	1	18	25	0	292	0	10000	5	2	1	12	55	0
239	1	95000	5	1.5	0	18	40	0	293	1	0	11	2	1	12	33	0
240	1	120000	15	1.5	1	0	40	0	294	1	69000	1	2	1	12	35	0
241	1	39000	20	0	1	18	40	0	295	1	70000	6	3.5	1	16	35	0
242	1	39000	10	1.5	1	18	40	1	296	1	69000	6	0	1	16	25	0
243	0	39000	11	1.5	1	18	40	1	297	1	70000	11	1.5	1	16	25	0
244	0	0	1	3	1	18	73	0	298	1	69000	11	1.5	1	16	65	1
245	1	35000	1	3	1	18	55	1	299	1	10000	7	1.5	1	16	47	0
246	1	39000	15	3	1	18	45	0	300	0	10000	1	1.5	1	16	45	1
247	0	70000	10	3	1	18	45	0	301	0	0	1	1.5	1	16	46	1
248	0	65000	21	3	1	18	77	0	302	1	10000	1	1.5	1	16	46	0
249	0	0	21	3	1	18	45	0	303	1	10000	17	1.5	1	16	46	0
250	0	70000	6	2	1	18	35	0	304	1	39000	10	1.5	1	16	46	1
251	1	40000	6	2	1	18	45	0	305	0	0	10	1.5	1	16	47	0
252	1	100000	1	2	1	18	45	1	306	0	0	10	3.5	1	18	25	0
253	1	0	1	1.5	1	18	45	0	307	1	39000	10	2.5	1	18	25	0
254	1	10000	6	1.5	1	12	45	0	308	1	0	10	2.5	1	6	25	0
255	1	10000	6	0	1	12	45	0	309	1	99000	20	3	1	0	25	0
256	1	10000	0	1.5	1	12	47	0	310	1	100000	5	3	1	6	27	0
257	1	10000	1	1.5	1	12	47	0	311	1	69000	15	3	1	6	27	1
258	1	10000	15	0	1	16	47	0	312	1	70000	15	3	0	6	27	0
259	1	65000	5	1.5	1	16	65	0	313	1	10000	10	3	1	6	33	0
260	1	0	10	1.5	1	16	35	0	314	1	10000	20	3.5	1	6	33	1
261	1	35000	10	1.5	1	16	35	1	315	1	99000	0	0	1	6	33	1
262	0	35000	10	1.5	1	16	44	0	316	1	0	15	0	1	6	35	1

Continuation.........

317	1	10000	5	0	0	6	35	0	370	1	65000	10	1.5	1	6	45	1
318	1	0	5	2	1	6	35	1	371	1	10000	11	1.5	1	12	49	0
319	0	10000	5	3.5	1	6	35	1	372	1	35000	11	0	1	12	47	0
320	1	39000	5	1.5	1	12	75	1	373	1	0	6	1.5	1	12	37	0
321	1	39000	10	1.5	1	12	35	1	374	1	35000	1	1.5	1	12	37	0
322	1	39000	15	1.5	1	12	35	0	375	0	35000	6	1.5	1	12	37	0
323	1	0	11	1.5	1	12	39	0	376	1	35000	1	1.5	1	12	44	0
324	1	100000	6	1.5	1	12	39	1	377	0	35000	1	1.5	0	12	47	0
325	1	100000	16	3.5	1	12	39	1	378	0	10000	16	1.5	1	16	47	0
326	0	69000	6	2.5	1	12	37	1	379	1	95000	6	1.5	1	16	49	1
327	0	100000	6	1.5	1	12	37	1	380	1	10000	6	1.5	1	16	49	1
328	1	40000	1	1.5	1	12	35	0	381	1	65000	1	1.5	1	16	47	1
329	1	100000	1	1.5	1	12	35	0	382	1	10000	6	1.5	1	16	47	1
330	0	70000	6	0	1	12	35	0	383	1	10000	16	1.5	1	16	47	0
331	1	100000	6	1.5	0	18	65	0	384	1	65000	1	0	1	16	57	0
332	1	69000	11	1.5	1	18	25	0	385	1	0	6	0	1	16	27	0
333	1	99000	1	1.5	1	18	25	1	386	1	95000	11	0	1	6	37	0
334	0	99000	1	1.5	1	18	25	1	387	1	65000	6	1.5	1	6	37	0
335	1	0	1	1.5	1	16	25	1	388	1	10000	6	1.5	1	6	37	1
336	1	69000	0	1.5	1	16	65	1	389	1	0	1	2	1	6	35	1
337	1	10000	1	4	1	16	65	1	390	1	15000	6	2	1	6	35	0
338	0	10000	6	1.5	1	16	25	1	391	1	15000	21	2	1	6	35	0
339	1	10000	10	1.5	1	16	25	0	392	1	100000	0	2	0	0	45	0
340	1	10000	10	1.5	1	16	55	0	393	1	65000	11	3.5	1	6	27	0
341	1	10000	11	1.5	1	16	37	0	394	1	100000	1	2.5	1	6	27	1
342	1	10000	1	1.5	1	16	35	0	395	1	65000	11	2.5	1	6	27	1
343	1	0	6	1.5	0	16	35	0	396	1	0	11	2.5	1	6	29	0
344	1	0	6	3.5	1	16	38	0	397	1	65000	6	0	1	6	29	0
345	1	0	11	2.5	1	6	39	0	398	1	70000	6	4	1	6	45	1
346	1	10000	6	0	1	6	47	1	399	1	10000	8	2.5	1	6	47	0
347	1	10000	6	0	1	20	44	1	400	1	0	9	2.5	1	6	47	0
348	1	35000	1	0	1	20	27	1	401	1	65000	6	2.5	0	16	47	0
349	0	35000	1	2	1	20	27	1	402	1	65000	11	2.5	0	16	47	0
350	0	11000	5	2	1	16	25	1	403	1	65000	10	2.5	1	16	47	0
351	0	70000	5	2	0	16	25	1	404	1	65000	10	2.5	1	16	47	0
352	1	10000	20	2	1	0	25	1	405	1	40000	10	2	1	16	55	0
353	1	0	5	2	1	16	27	0	406	1	40000	5	2	1	16	37	0
354	1	0	15	1.5	1	16	28	0	407	1	35000	20	2	1	16	38	0
355	1	65000	15	3.5	1	16	27	1	408	1	35000	10	2	1	16	39	0
356	1	15000	20	0	1	16	45	1	409	1	35000	10	2	1	16	40	0
357	1	10000	7	3	1	6	45	1	410	0	35000	5	3.5	1	16	45	0
358	1	95000	1	3	1	6	47	0	411	1	35000	5	2.5	1	16	43	1
359	1	10000	15	3	1	6	47	0	412	0	10000	10	2.5	1	16	39	1
360	1	65000	10	3	1	6	47	0	413	1	95000	10	2.5	1	16	29	1
361	1	100000	20	3	1	6	47	1	414	1	100000	10	2.5	1	16	27	1
362	1	10000	10	2.5	1	6	49	1	415	1	65000	5	2.5	1	16	28	1
363	1	65000	10	2.5	1	6	49	0	416	1	40000	11	3.5	1	16	28	1
364	1	10000	0	4	1	6	77	0	417	1	0	10	1.5	1	18	33	0
365	1	95000	5	1.5	1	6	29	0	418	0	95000	15	0	1	18	33	0
366	1	0	5	3.5	0	6	29	1	419	1	100000	5	1.5	1	18	37	0
367	1	65000	5	2.5	1	6	75	0	420	1	0	5	1.5	0	12	39	0
368	1	10000	20	4	1	6	45	0	421	1	0	10	1.5	1	12	37	0
369	1	10000	10	1.5	1	6	45	1	422	1	95000	15	1.5	1	12	37	0

Continuation..........

423	1	95000	15	1.5	1	12	39	0	478	1	35000	15	2.5	0	20	37	1
424	1	69000	10	1.5	1	12	47	1	479	1	35000	11	2.5	0	20	37	0
425	1	0	0	1.5	1	12	47	1	480	1	95000	1	2.5	0	12	37	1
426	1	10000	15	1.5	1	16	49	1	481	1	0	11	2.5	1	12	37	1
427	1	0	5	4	1	16	49	1	482	1	100000	11	2.5	1	12	65	0
428	1	10000	5	3.5	1	16	50	1	483	1	100000	11	2.5	1	12	27	1
429	1	10000	15	3.5	1	0	55	1	484	1	100000	5	2.5	1	12	55	1
430	1	10000	11	2	0	16	75	1	485	1	0	15	0	1	12	45	0
431	0	100000	6	2	1	16	67	0	486	1	100000	15	2	1	12	45	0
432	1	69000	1	2	1	20	47	0	487	0	100000	20	2	1	16	45	0
433	1	100000	11	2	1	20	47	0	488	1	10000	10	2	1	16	47	1
434	1	100000	11	2	1	20	27	0	489	1	15000	10	2	1	16	49	1
435	1	99000	11	0	1	16	25	0	490	1	15000	5	2	1	16	44	1
436	1	100000	1	0	1	16	25	0	491	1	15000	20	2	1	16	25	0
437	1	40000	11	3.5	1	16	25	0	492	1	0	5	2	1	16	25	0
438	1	100000	6	3.5	1	16	28	0	493	1	100000	20	3.5	1	16	55	1
439	1	70000	6	3	1	16	28	0	494	1	99000	5	2.5	1	16	55	1
440	1	100000	1	0	1	16	29	0	495	1	100000	5	3.5	1	0	35	0
441	0	0	1	3	1	16	29	0	496	1	100000	5	3.5	1	16	35	0
442	1	39000	11	4	1	16	45	1	497	1	100000	20	3.5	1	16	35	1
443	1	39000	6	1.5	0	16	47	1	498	0	100000	20	3.5	1	16	37	1
444	1	39000	6	1.5	1	16	49	1	499	1	70000	5	3.5	1	16	37	1
445	1	39000	1	1.5	1	16	49	1	500	1	100000	5	3.5	1	16	60	1
446	1	39000	16	1.5	1	16	49	1	501	1	100000	20	0	1	20	29	0
447	1	39000	6	1.5	1	12	49	1	502	1	100000	20	3.5	1	20	29	0
448	1	99000	1	1.5	1	12	45	1	503	1	95000	5	3.5	1	20	65	0
449	1	100000	11	1.5	1	12	45	1	504	1	100000	5	3.5	1	20	44	0
450	1	69000	11	0	1	12	27	1	505	1	100000	10	0	1	16	44	0
451	0	100000	11	0	1	12	27	0	506	1	35000	10	2.5	1	16	47	0
452	1	0	11	1.5	1	12	27	0	507	1	35000	10	2.5	1	16	48	0
453	1	0	1	4	1	12	29	0	508	1	0	5	2.5	0	16	33	0
454	1	0	11	3.5	1	18	37	0	509	1	35000	5	2.5	1	12	37	0
455	1	10000	11	3.5	1	18	37	0	510	1	35000	5	2.5	1	12	35	1
456	1	10000	6	2	1	18	37	0	511	1	35000	1	2.5	1	12	70	0
457	1	10000	16	2	1	18	37	0	512	1	0	1	2.5	1	12	27	1
458	1	65000	1	2	1	18	29	0	513	1	35000	1	2.5	1	12	37	0
459	1	95000	11	2	1	18	29	1	514	1	15000	1	2.5	1	12	35	0
460	0	10000	11	2	1	18	65	0	515	0	15000	1	2.5	1	12	35	0
461	1	10000	1	2	1	18	55	0	516	1	15000	20	2.5	1	12	35	0
462	1	0	11	0	1	18	45	1	517	1	15000	5	3.5	1	12	35	0
463	1	10000	6	3.5	1	0	55	1	518	1	95000	20	3.5	1	12	37	0
464	1	10000	6	1.5	1	16	35	0	519	1	100000	21	1.5	1	12	37	0
465	1	95000	1	1.5	1	16	37	0	520	1	100000	16	1.5	1	18	37	1
466	1	65000	11	1.5	1	16	44	0	521	1	100000	3	1.5	1	18	70	1
467	1	65000	0	1.5	1	16	45	0	522	1	0	3	4	1	18	50	1
468	1	100000	1	1.5	1	16	44	0	523	1	100000	3	3.5	1	18	69	1
469	1	100000	6	1.5	1	16	46	1	524	1	95000	16	0	1	20	33	1
470	1	95000	6	1.5	1	16	47	1	525	1	100000	6	4	1	20	33	1
471	1	95000	6	1.5	1	16	47	1	526	1	100000	1	4	1	20	27	0
472	1	0	6	1.5	1	16	46	0	527	1	100000	21	4	1	12	73	0
473	1	100000	10	1.5	1	16	50	1	528	1	100000	6	0	1	12	45	0
474	0	35000	5	3.5	1	16	27	1	529	1	100000	6	2	1	12	47	0
475	1	35000	15	0	1	16	29	1	530	1	100000	1	2	1	6	47	0
476	1	35000	15	0	1	20	69	1	531	1	99000	1	2	1	6	47	0
477	1	35000	15	2.5	1	20	37	0	532	1	99000	5	2	0	6	50	0

Continuation.........

533	1	100000	5	2	1	6	55	0
534	1	0	20	2	1	6	73	0
535	1	100000	5	2	1	6	27	1
536	1	100000	21	4	1	6	27	1
537	1	100000	6	4	1	6	33	1
538	1	100000	0	1.5	1	6	33	1
539	1	95000	7	1.5	1	6	37	1
540	1	95000	21	1.5	1	6	35	1
541	1	95000	6	1.5	1	12	36	1
542	1	100000	6	1.5	1	16	36	0
543	1	100000	6	1.5	1	16	66	1
544	1	95000	7	1.5	0	16	44	1
545	1	95000	7	4	1	16	43	1
546	1	100000	16	4	1	16	44	1
547	1	0	7	2	1	16	50	0
548	1	0	7	0	1	16	49	1
549	1	35000	7	1.5	0	12	45	1
p550	1	35000	21	0	1	18	45	1

Source: Field Survey, 2010.

APPENDIX II

Estimated Logit Regression Equation

Dependent Variable: HFINS
Method: ML - Binary Logit
Date: 02/27/11   Time: 23:05
Sample: 1 550
Included observations: 550
Convergence achieved after 10 iterations
Covariance matrix computed using second derivatives

Variable	Coefficient	Std. Error	z-Statistic	Prob.
C	3.358324	0.909525	3.692394	0.0002
HINC	7.45E-08	3.07E-06	0.024291	0.9806
HFSZ	0.014256	0.019197	0.742610	0.4577
HFMS	-0.142075	0.111753	-1.271331	0.2036
HSEX	-0.809833	0.622452	-1.301037	0.1932
HEDU	-0.081234	0.029904	-2.716441	0.0066
HAGE	0.018014	0.010924	1.649035	0.0991
HEFM	-0.609177	0.271044	-2.247518	0.0246

Mean dependent var	0.821818	S.D. dependent var	0.383014
S.E. of regression	0.378876	Akaike info criterion	0.929750
Sum squared resid	77.80233	Schwarz criterion	0.992440
Log likelihood	-247.6812	Hannan-Quinn criter.	0.954248
Restr. log likelihood	-257.7439	Avg. log likelihood	-0.450329
LR statistic (7 df)	20.12537	McFadden R-squared	0.039041
Probability(LR stat)	0.005305		

Obs with Dep=0	98	Total obs	550
Obs with Dep=1	452		

APPENDIX III

Andrews and Hosmer-Lemeshow Goodness-of-Fit Tests

Dependent Variable: HFINS
Method: ML - Binary Logit
Date: 02/27/11   Time: 23:05
Sample: 1 550
Included observations: 550
Andrews and Hosmer-Lemeshow Goodness-of-Fit Tests
Grouping based upon predicted risk (randomize ties)

	Quantile of Risk		Dep=0		Dep=1		Total	H-L
	Low	High	Actual	Expect	Actual	Expect	Obs	Value
1	0.5633	0.7212	15	17.1237	40	37.8763	55	0.38245
2	0.7217	0.7558	15	14.3565	40	40.6435	55	0.03904
3	0.7559	0.7848	16	12.6244	39	42.3756	55	1.17150
4	0.7848	0.8080	14	11.1178	41	43.8822	55	0.93646
5	0.8087	0.8288	6	9.93013	49	45.0699	55	1.89817
6	0.8291	0.8480	8	8.89396	47	46.1060	55	0.10719
7	0.8485	0.8649	8	7.90090	47	47.0991	55	0.00145
8	0.8653	0.8861	7	6.81351	48	48.1865	55	0.00583
9	0.8863	0.9078	5	5.60638	50	49.3936	55	0.07303
10	0.9085	1.0000	4	3.63274	51	51.3673	55	0.03975
	Total		98	98.0000	452	452.000	550	4.65487

H-L Statistic:	4.6549		Prob[Chi-Sq(8 df)]:	0.7937
Andrews Statistic:	5.1950		Prob[Chi-Sq(10 df)]:	0.8778

APPENDIX IV

ACTUAL, FITTED AND RESIDUAL GRAPH PLOTS

```
1.00000   0.81243    0.18757          |    .  |*  .   |
1.00000   0.91131    0.08869          |    .  |*  .   |
1.00000   0.94721    0.05279          |    .  |*  .   |
1.00000   0.75827    0.24173          |    .  | *  .  |
1.00000   0.82393    0.17607          |    .  | *  .  |
1.00000   0.74189    0.25811          |    .  | *  .  |
1.00000   0.84091    0.15909          |    .  | *  .  |
1.00000   0.81853    0.18147          |    .  | *  .  |
1.00000   0.82877    0.17123          |    .  | *  .  |
1.00000   0.91684    0.08316          |    .  |*  .   |
1.00000   0.92110    0.07890          |    .  |*  .   |
1.00000   0.81856    0.18144          |    .  | *  .  |
1.00000   0.82077    0.17923          |    .  | *  .  |
1.00000   0.86380    0.13620          |    .  | *  .  |
1.00000   0.87222    0.12778          |    .  | *  .  |
1.00000   0.85059    0.14941          |    .  | *  .  |
0.00000   0.86424   -0.86424          | *  .  |    .  |
1.00000   0.79022    0.20978          |    .  | *  .  |
1.00000   0.89765    0.10235          |    .  |*  .   |
1.00000   0.76790    0.23210          |    .  | *  .  |
1.00000   0.93807    0.06193          |    .  |*  .   |
1.00000   0.76547    0.23453          |    .  | *  .  |
1.00000   0.78418    0.21582          |    .  | *  .  |
1.00000   0.72369    0.27631          |    .  | *  .  |
1.00000   0.67959    0.32041          |    .  | *  .  |
0.00000   0.69999   -0.69999          |   *  .|    .  |
1.00000   0.78482    0.21518          |    .  | *  .  |
1.00000   0.77258    0.22742          |    .  | *  .  |
1.00000   0.80870    0.19130          |    .  | *  .  |
1.00000   0.85711    0.14289          |    .  | *  .  |
1.00000   0.85341    0.14659          |    .  | *  .  |
0.00000   0.83675   -0.83675          | *  .  |    .  |
0.00000   0.74941   -0.74941          | *  .  |    .  |
0.00000   0.87972   -0.87972          | *  .  |    .  |
1.00000   0.85519    0.14481          |    .  | *  .  |
1.00000   0.90669    0.09331          |    .  |*  .   |
1.00000   0.87200    0.12800          |    .  | *  .  |
1.00000   0.83900    0.16100          |    .  | *  .  |
1.00000   0.82882    0.17118          |    .  | *  .  |
1.00000   0.85545    0.14455          |    .  | *  .  |
1.00000   0.72490    0.27510          |    .  | *  .  |
1.00000   0.72490    0.27510          |    .  | *  .  |
1.00000   0.80786    0.19214          |    .  | *  .  |
1.00000   0.80786    0.19214          |    .  | *  .  |
1.00000   0.87996    0.12004          |    .  | *  .  |
1.00000   0.81052    0.18948          |    .  | *  .  |
1.00000   0.80431    0.19569          |    .  | *  .  |
1.00000   0.89701    0.10299          |    .  |*  .   |
1.00000   0.90348    0.09652          |    .  |*  .   |
0.00000   0.78943   -0.78943          |  *   .|    .  |
0.00000   0.77786   -0.77786          |  *   .|    .  |
1.00000   0.86557    0.13443          |    .  | *  .  |
1.00000   0.80092    0.19908          |    .  | *  .  |
1.00000   0.80104    0.19896          |    .  | *  .  |
1.00000   0.87642    0.12358          |    .  | *  .  |
1.00000   0.94471    0.05529          |    .  |*  .   |
```

```
1.00000    0.95348    0.04652     |      .    |* .    |
1.00000    0.80557    0.19443     |      .    |* .    |
1.00000    0.89206    0.10794     |      .    |* .    |
0.00000    0.90583   -0.90583     |*     .    |  .    |
0.00000    0.91877   -0.91877     |*     .    |  .    |
1.00000    0.90077    0.09923     |      .    |* .    |
1.00000    0.90129    0.09871     |      .    |* .    |
1.00000    0.92498    0.07502     |      .    |* .    |
1.00000    0.86197    0.13803     |      .    |* .    |
1.00000    0.86295    0.13705     |      .    |* .    |
1.00000    0.90578    0.09422     |      .    |* .    |
0.00000    0.89573   -0.89573     |*     .    |  .    |
0.00000    0.88194   -0.88194     |  *   .    |  .    |
1.00000    0.87300    0.12700     |      .    |* .    |
1.00000    0.89417    0.10583     |      .    |* .    |
1.00000    0.90092    0.09908     |      .    |* .    |
1.00000    0.96578    0.03422     |      .    | *.    |
1.00000    0.91006    0.08994     |      .    |* .    |
1.00000    0.88703    0.11297     |      .    |* .    |
1.00000    0.77608    0.22392     |      .    |  *.   |
0.00000    0.76345   -0.76345     |   *  .    |  .    |
0.00000    0.77568   -0.77568     |   *  .    |  .    |
0.00000    0.89255   -0.89255     | *    .    |  .    |
0.00000    0.89845   -0.89845     |*     .    |  .    |
0.00000    0.78483   -0.78483     |  *   .    |  .    |
0.00000    0.78357   -0.78357     |  *   .    |  .    |
1.00000    0.71790    0.28210     |      .    |   *.  |
1.00000    0.74926    0.25074     |      .    |  *.   |
1.00000    0.87228    0.12772     |      .    |* .    |
1.00000    0.93885    0.06115     |      .    |* .    |
1.00000    0.73089    0.26911     |      .    |   *.  |
1.00000    0.77149    0.22851     |      .    |  *.   |
1.00000    0.81785    0.18215     |      .    |  *.   |
1.00000    0.80699    0.19301     |      .    |  *.   |
1.00000    0.70934    0.29066     |      .    |    *. |
1.00000    0.96437    0.03563     |      .    | *.    |
1.00000    0.81774    0.18226     |      .    |  *.   |
0.00000    0.67466   -0.67466     |    * .    |  .    |
0.00000    0.71643   -0.71643     |    * .    |  .    |
1.00000    0.82829    0.17171     |      .    |  *.   |
1.00000    0.86134    0.13866     |      .    |* .    |
1.00000    0.93750    0.06250     |      .    |* .    |
1.00000    0.70947    0.29053     |      .    |    *. |
1.00000    0.75191    0.24809     |      .    |  *.   |
1.00000    0.82856    0.17144     |      .    |  *.   |
1.00000    0.83237    0.16763     |      .    |  *.   |
1.00000    0.86095    0.13905     |      .    |* .    |
1.00000    0.84309    0.15691     |      .    |  *.   |
1.00000    0.75917    0.24083     |      .    |  *.   |
0.00000    0.82430   -0.82430     | *    .    |  .    |
0.00000    0.73260   -0.73260     |   *  .    |  .    |
1.00000    0.92870    0.07130     |      .    |* .    |
1.00000    0.84804    0.15196     |      .    |  *.   |
1.00000    0.76481    0.23519     |      .    |  *.   |
1.00000    0.77739    0.22261     |      .    |  *.   |
1.00000    0.89377    0.10623     |      .    |* .    |
1.00000    0.76211    0.23789     |      .    |  *.   |
1.00000    0.82147    0.17853     |      .    |  *.   |
1.00000    0.82545    0.17455     |      .    |  *.   |
0.00000    0.78227   -0.78227     |  *   .    |  .    |
```

```
0.00000    0.92886    -0.92886        |*     .    |    .    |
0.00000    0.83121    -0.83121        | *    .    |    .    |
0.00000    0.74414    -0.74414        |  *   .    |    .    |
1.00000    0.72988     0.27012        |      .    |  * .    |
1.00000    0.88414     0.11586        |      .    |*  .    |
1.00000    0.88414     0.11586        |      .    |*  .    |
1.00000    0.87583     0.12417        |      .    | * .    |
1.00000    0.79367     0.20633        |      .    |  *.    |
1.00000    0.84068     0.15932        |      .    | * .    |
1.00000    0.86729     0.13271        |      .    | * .    |
1.00000    0.86244     0.13756        |      .    | * .    |
1.00000    0.68390     0.31610        |      .    |   *.    |
0.00000    0.72850    -0.72850        |    * .    |    .    |
0.00000    0.72850    -0.72850        |    * .    |    .    |
0.00000    0.84195    -0.84195        |*     .    |    .    |
0.00000    0.74181    -0.74181        |   *  .    |    .    |
0.00000    0.74268    -0.74268        |   *  .    |    .    |
1.00000    0.81093     0.18907        |      .    |  *.    |
1.00000    0.83681     0.16319        |      .    | * .    |
1.00000    0.89776     0.10224        |      .    |*  .    |
1.00000    0.83641     0.16359        |      .    | * .    |
1.00000    0.85063     0.14937        |      .    | * .    |
1.00000    0.85054     0.14946        |      .    | * .    |
1.00000    0.80438     0.19562        |      .    | * .    |
1.00000    0.89843     0.10157        |      .    |*  .    |
1.00000    0.81536     0.18464        |      .    | * .    |
1.00000    0.72125     0.27875        |      .    |  * .    |
1.00000    0.72125     0.27875        |      .    |  *.    |
1.00000    0.93805     0.06195        |      .    |*  .    |
1.00000    0.78759     0.21241        |      .    |  *.    |
1.00000    0.74934     0.25066        |      .    |  *.    |
1.00000    0.74934     0.25066        |      .    |  *.    |
1.00000    0.84571     0.15429        |      .    | * .    |
0.00000    0.85691    -0.85691        |*     .    |    .    |
0.00000    0.86351    -0.86351        |*     .    |    .    |
1.00000    0.75950     0.24050        |      .    |  *.    |
1.00000    0.77128     0.22872        |      .    |  *.    |
1.00000    0.87831     0.12169        |      .    | * .    |
1.00000    0.70097     0.29903        |      .    |   *.    |
1.00000    0.70474     0.29526        |      .    |   *.    |
1.00000    0.70722     0.29278        |      .    |   *.    |
1.00000    0.81547     0.18453        |      .    |  *.    |
1.00000    0.76179     0.23821        |      .    |  *.    |
1.00000    0.81007     0.18993        |      .    |  *.    |
1.00000    0.77480     0.22520        |      .    |  * .    |
0.00000    0.72825    -0.72825        |    * .    |    .    |
0.00000    0.83116    -0.83116        |*     .    |    .    |
1.00000    0.82911     0.17089        |      .    | * .    |
1.00000    0.82972     0.17028        |      .    | * .    |
1.00000    0.80748     0.19252        |      .    | * .    |
1.00000    0.83149     0.16851        |      .    | * .    |
1.00000    0.82304     0.17696        |      .    | * .    |
1.00000    0.76709     0.23291        |      .    |  *.    |
0.00000    0.77842    -0.77842        |   *  .    |    .    |
0.00000    0.72941    -0.72941        |    * .    |    .    |
1.00000    0.90459     0.09541        |      .    |*  .    |
1.00000    0.74781     0.25219        |      .    |  *.    |
1.00000    0.81106     0.18894        |      .    |  *.    |
1.00000    0.72013     0.27987        |      .    |  *.    |
1.00000    0.73704     0.26296        |      .    |  *.    |
```

86

```
1.00000   0.73968    0.26032        |    .   |  * .   |
1.00000   0.79814    0.20186        |    .   |  * .   |
1.00000   0.73717    0.26283        |    .   |  * .   |
1.00000   0.72752    0.27248        |    .   |  * .   |
1.00000   0.72641    0.27359        |    .   |  * .   |
1.00000   0.74077    0.25923        |    .   |  * .   |
1.00000   0.93868    0.06132        |    .   |* .     |
1.00000   0.86470    0.13530        |    .   | * .    |
1.00000   0.81262    0.18738        |    .   | * .    |
1.00000   0.66464    0.33536        |    .   |   * .  |
0.00000   0.68732   -0.68732        |  * .   |    .   |
0.00000   0.81248   -0.81248        | * .    |    .   |
0.00000   0.68360   -0.68360        |  * .   |    .   |
0.00000   0.69581   -0.69581        |  * .   |    .   |
0.00000   0.74031   -0.74031        |  * .   |    .   |
1.00000   0.69638    0.30362        |    .   |   *.   |
1.00000   0.73992    0.26008        |    .   |  * .   |
1.00000   0.83853    0.16147        |    .   | * .    |
1.00000   0.83211    0.16789        |    .   | * .    |
1.00000   0.84363    0.15637        |    .   | * .    |
1.00000   0.93793    0.06207        |    .   |* .     |
1.00000   0.76902    0.23098        |    .   |  * .   |
0.00000   0.85526   -0.85526        | * .    |    .   |
0.00000   0.79718   -0.79718        | * .    |    .   |
1.00000   0.86994    0.13006        |    .   | * .    |
1.00000   0.87819    0.12181        |    .   | * .    |
1.00000   0.85508    0.14492        |    .   | * .    |
1.00000   0.85536    0.14464        |    .   | * .    |
1.00000   0.88946    0.11054        |    .   |* .     |
0.00000   0.79669   -0.79669        | * .    |    .   |
0.00000   0.82914   -0.82914        | * .    |    .   |
1.00000   0.80799    0.19201        |    .   | * .    |
1.00000   0.73663    0.26337        |    .   |  * .   |
1.00000   0.86489    0.13511        |    .   | * .    |
1.00000   0.84899    0.15101        |    .   | * .    |
1.00000   0.87174    0.12826        |    .   | * .    |
1.00000   0.70993    0.29007        |    .   |   *.   |
1.00000   0.79838    0.20162        |    .   |  * .   |
1.00000   0.74504    0.25496        |    .   |  * .   |
1.00000   0.66385    0.33615        |    .   |   * .  |
0.00000   0.67524   -0.67524        |  * .   |    .   |
0.00000   0.71058   -0.71058        |  * .   |    .   |
1.00000   0.81900    0.18100        |    .   | * .    |
1.00000   0.85667    0.14333        |    .   | * .    |
1.00000   0.85234    0.14766        |    .   | * .    |
1.00000   0.88314    0.11686        |    .   | * .    |
1.00000   0.90240    0.09760        |    .   |* .     |
1.00000   0.86396    0.13604        |    .   | * .    |
0.00000   0.85075   -0.85075        | * .    |    .   |
0.00000   0.68689   -0.68689        |  * .   |    .   |
0.00000   0.88606   -0.88606        | * .    |    .   |
0.00000   0.87215   -0.87215        | * .    |    .   |
0.00000   0.86791   -0.86791        | * .    |    .   |
0.00000   0.85952   -0.85952        | * .    |    .   |
1.00000   0.89440    0.10560        |    .   |* .     |
1.00000   0.83246    0.16754        |    .   | * .    |
1.00000   0.73692    0.26308        |    .   |  * .   |
1.00000   0.85957    0.14043        |    .   | * .    |
1.00000   0.88041    0.11959        |    .   | * .    |
1.00000   0.90783    0.09217        |    .   |* .     |
```

```
1.00000   0.70201    0.29799      |     .   |  *. |
1.00000   0.71724    0.28276      |     .   |  *. |
1.00000   0.86678    0.13322      |     .   | * . |
1.00000   0.93521    0.06479      |     .   |*  . |
1.00000   0.81543    0.18457      |     .   | *  . |
0.00000   0.75584   -0.75584      | *   .   |    . |
0.00000   0.75846   -0.75846      | *   .   |    . |
1.00000   0.87967    0.12033      |     .   | *. |
1.00000   0.74241    0.25759      |     .   |  *. |
0.00000   0.74616   -0.74616      | *   .   |    . |
0.00000   0.73288   -0.73288      |  *  .   |    . |
0.00000   0.85096   -0.85096      | *   .   |    . |
0.00000   0.76149   -0.76149      | *   .   |    . |
1.00000   0.71386    0.28614      |     .   |  *. |
1.00000   0.74878    0.25122      |     .   |  * . |
1.00000   0.73600    0.26400      |     .   |  *. |
1.00000   0.84528    0.15472      |     .   | * . |
1.00000   0.90529    0.09471      |     .   |* . |
1.00000   0.92205    0.07795      |     .   |* . |
1.00000   0.90096    0.09904      |     .   |* . |
1.00000   0.90223    0.09777      |     .   |* . |
1.00000   0.90970    0.09030      |     .   |* . |
1.00000   0.90743    0.09257      |     .   |* . |
1.00000   0.85920    0.14080      |     .   | * . |
0.00000   0.76890   -0.76890      | *   .   |    . |
0.00000   0.79645   -0.79645      | *   .   |    . |
0.00000   0.79413   -0.79413      | *   .   |    . |
0.00000   0.79413   -0.79413      | *   .   |    . |
0.00000   0.76025   -0.76025      | *   .   |    . |
0.00000   0.82673   -0.82673      | *   .   |    . |
1.00000   0.75656    0.24344      |     .   |  *. |
1.00000   0.87622    0.12378      |     .   | * . |
1.00000   0.84182    0.15818      |     .   | * . |
1.00000   0.78741    0.21259      |     .   | *. |
1.00000   0.78809    0.21191      |     .   | *. |
1.00000   0.93455    0.06545      |     .   |* . |
0.00000   0.79334   -0.79334      | *   .   |    . |
0.00000   0.70642   -0.70642      | *   .   |    . |
1.00000   0.79916    0.20084      |     .   | *. |
1.00000   0.76196    0.23804      |     .   |  *. |
1.00000   0.73558    0.26442      |     .   |  *. |
1.00000   0.88338    0.11662      |     .   | *. |
1.00000   0.79321    0.20679      |     .   | *. |
1.00000   0.82621    0.17379      |     .   | *. |
1.00000   0.93597    0.06403      |     .   |* . |
1.00000   0.82129    0.17871      |     .   | *. |
1.00000   0.84679    0.15321      |     .   | * . |
1.00000   0.76874    0.23126      |     .   |  *. |
1.00000   0.85474    0.14526      |     .   | * . |
1.00000   0.90651    0.09349      |     .   |* . |
1.00000   0.94538    0.05462      |     .   |* . |
1.00000   0.92606    0.07394      |     .   |* . |
1.00000   0.83900    0.16100      |     .   | * . |
1.00000   0.82675    0.17325      |     .   | *. |
0.00000   0.77531   -0.77531      | *   .   |    . |
1.00000   0.85109    0.14891      |     .   | *. |
1.00000   0.88502    0.11498      |     .   | *. |
1.00000   0.87429    0.12571      |     .   | *. |
1.00000   0.81347    0.18653      |     .   | *. |
1.00000   0.85690    0.14310      |     .   | *. |
```

```
1.00000   0.73863    0.26137      |        .    |  * .    |
1.00000   0.85313    0.14687      |        .    |  * .    |
0.00000   0.79797   -0.79797      |  *     .    |    .    |
0.00000   0.77766   -0.77766      |  *     .    |    .    |
1.00000   0.78063    0.21937      |        .    |   *.    |
1.00000   0.86753    0.13247      |        .    |  * .    |
1.00000   0.89162    0.10838      |        .    | *.     |
0.00000   0.80227   -0.80227      |  *     .    |    .    |
0.00000   0.80466   -0.80466      |  *     .    |    .    |
1.00000   0.63940    0.36060      |        .    |    . *  |
1.00000   0.67211    0.32789      |        .    |    .*.  |
1.00000   0.84418    0.15582      |        .    |  * .    |
0.00000   0.90516   -0.90516      |*       .    |    .    |
1.00000   0.83073    0.16927      |        .    |  * .    |
1.00000   0.84955    0.15045      |        .    |  * .    |
1.00000   0.95891    0.04109      |        .    | *.     |
1.00000   0.91472    0.08528      |        .    | *.     |
1.00000   0.86235    0.13765      |        .    |  * .    |
1.00000   0.88633    0.11367      |        .    | *.     |
1.00000   0.90857    0.09143      |        .    | *.     |
1.00000   0.97271    0.02729      |        .    | * .    |
0.00000   0.86640   -0.86640      |  *     .    |    .    |
1.00000   0.83986    0.16014      |        .    |  * .    |
1.00000   0.89813    0.10187      |        .    | *.     |
1.00000   0.82162    0.17838      |        .    |  * .    |
1.00000   0.90095    0.09905      |        .    | *.     |
1.00000   0.90203    0.09797      |        .    | *.     |
1.00000   0.82447    0.17553      |        .    |  * .    |
0.00000   0.80303   -0.80303      |  *     .    |    .    |
0.00000   0.79682   -0.79682      |  *     .    |    .    |
1.00000   0.81920    0.18080      |        .    |  * .    |
1.00000   0.88167    0.11833      |        .    |  * .    |
0.00000   0.88213   -0.88213      |  *     .    |    .    |
1.00000   0.90846    0.09154      |        .    | *.     |
1.00000   0.95012    0.04988      |        .    | *.     |
1.00000   0.81540    0.18460      |        .    |  * .    |
0.00000   0.67612   -0.67612      |     *  .    |    .    |
1.00000   0.67612    0.32388      |        .    |    .*.  |
1.00000   0.70911    0.29089      |        .    |    .*.  |
1.00000   0.83237    0.16763      |        .    |  * .    |
0.00000   0.77854   -0.77854      |  *     .    |    .    |
1.00000   0.72374    0.27626      |        .    |   *.    |
1.00000   0.83607    0.16393      |        .    |  * .    |
1.00000   0.89750    0.10250      |        .    | *.     |
1.00000   0.86526    0.13474      |        .    |  * .    |
1.00000   0.84305    0.15695      |        .    |  * .    |
1.00000   0.92834    0.07166      |        .    | *.     |
1.00000   0.82077    0.17923      |        .    |  * .    |
1.00000   0.92859    0.07141      |        .    | *.     |
1.00000   0.91566    0.08434      |        .    | *.     |
1.00000   0.76737    0.23263      |        .    |   *.    |
0.00000   0.69379   -0.69379      |    *   .    |    .    |
0.00000   0.63035   -0.63035      |      * .    |    .    |
0.00000   0.70638   -0.70638      |    *   .    |    .    |
1.00000   0.84450    0.15550      |        .    |  * .    |
1.00000   0.91617    0.08383      |        .    | *.     |
1.00000   0.82087    0.17913      |        .    |  * .    |
1.00000   0.85244    0.14756      |        .    |  * .    |
1.00000   0.81100    0.18900      |        .    |  * .    |
1.00000   0.85023    0.14977      |        .    |  * .    |
```

```
1.00000    0.87401    0.12599        |      .    |*  .    |
1.00000    0.86917    0.13083        |      .    |*  .    |
1.00000    0.88963    0.11037        |      .    |*  .    |
1.00000    0.88285    0.11715        |      .    |*  .    |
1.00000    0.89705    0.10295        |      .    |*  .    |
1.00000    0.89309    0.10691        |      .    |*  .    |
1.00000    0.93912    0.06088        |      .    |*  .    |
1.00000    0.94688    0.05312        |      .    |*  .    |
1.00000    0.92047    0.07953        |      .    |*  .    |
1.00000    0.91358    0.08642        |      .    |*  .    |
1.00000    0.92580    0.07420        |      .    |*  .    |
1.00000    0.87869    0.12131        |      .    |*  .    |
1.00000    0.89960    0.10040        |      .    |*  .    |
1.00000    0.89996    0.10004        |      .    |*  .    |
1.00000    0.85713    0.14287        |      .    |*  .    |
1.00000    0.87767    0.12233        |      .    |*  .    |
1.00000    0.81809    0.18191        |      .    |*  .    |
0.00000    0.80764   -0.80764        | *    .    |   .    |
0.00000    0.81848   -0.81848        | *    .    |   .    |
0.00000    0.82648   -0.82648        |*     .    |   .    |
0.00000    0.91870   -0.91870        |*     .    |   .    |
1.00000    0.81787    0.18213        |      .    |*  .    |
1.00000    0.80246    0.19754        |      .    |  *.    |
1.00000    0.80146    0.19854        |      .    |  *.    |
1.00000    0.78452    0.21548        |      .    |   *.   |
1.00000    0.79566    0.20434        |      .    |  *.    |
1.00000    0.81787    0.18213        |      .    |*  .    |
1.00000    0.84363    0.15637        |      .    |*  .    |
1.00000    0.77056    0.22944        |      .    |   *    |
1.00000    0.90740    0.09260        |      .    |*  .    |
1.00000    0.88034    0.11966        |      .    |*  .    |
1.00000    0.87991    0.12009        |      .    |*  .    |
1.00000    0.85966    0.14034        |      .    |*  .    |
1.00000    0.86817    0.13183        |      .    |*  .    |
1.00000    0.89078    0.10922        |      .    |*  .    |
1.00000    0.96384    0.03616        |      .    |*  .    |
1.00000    0.83239    0.16761        |      .    |*  .    |
1.00000    0.83269    0.16731        |      .    |*  .    |
1.00000    0.85129    0.14871        |      .    |*  .    |
1.00000    0.85520    0.14480        |      .    |*  .    |
1.00000    0.88742    0.11258        |      .    |*  .    |
1.00000    0.85631    0.14369        |      .    |*  .    |
1.00000    0.93507    0.06493        |      .    |*  .    |
1.00000    0.93589    0.06411        |      .    |*  .    |
1.00000    0.93342    0.06658        |      .    |*  .    |
1.00000    0.93771    0.06229        |      .    |*  .    |
1.00000    0.86848    0.13152        |      .    |*  .    |
1.00000    0.86848    0.13152        |      .    |*  .    |
1.00000    0.89099    0.10901        |      .    |*  .    |
1.00000    0.84624    0.15376        |      .    |*  .    |
1.00000    0.87401    0.12599        |      .    |*  .    |
1.00000    0.85964    0.14036        |      .    |*  .    |
0.00000    0.86180   -0.86180        |*     .    |   .    |
0.00000    0.83699   -0.83699        |*     .    |   .    |
0.00000    0.75637   -0.75637        | *    .    |   .    |
1.00000    0.75588    0.24412        |      .    |   *.   |
1.00000    0.72241    0.27759        |      .    |    *.  |
1.00000    0.71520    0.28480        |      .    |    *.  |
1.00000    0.70368    0.29632        |      .    |    *.  |
1.00000    0.69136    0.30864        |      .    |    *.  |
```

90

```
0.00000   0.73126   -0.73126    |*    .    |    .    |
1.00000   0.78457    0.21543    |    .    |*   .    |
1.00000   0.73287    0.26713    |    .    |*   .    |
1.00000   0.91174    0.08826    |    .    |*   .    |
1.00000   0.82642    0.17358    |    .    |*   .    |
1.00000   0.83738    0.16262    |    .    |*   .    |
1.00000   0.84222    0.15778    |    .    |*   .    |
1.00000   0.85142    0.14858    |    .    |*   .    |
1.00000   0.83175    0.16825    |    .    |*   .    |
1.00000   0.82109    0.17891    |    .    |*   .    |
1.00000   0.73599    0.26401    |    .    |*   .    |
1.00000   0.75307    0.24693    |    .    |*   .    |
1.00000   0.93385    0.06615    |    .    |*   .    |
0.00000   0.93547   -0.93547    |*    .    |    .    |
1.00000   0.83961    0.16039    |    .    |*   .    |
1.00000   0.71024    0.28976    |    .    |   *.    |
1.00000   0.73912    0.26088    |    .    |   *.    |
1.00000   0.66399    0.33601    |    .    |    *.   |
1.00000   0.77801    0.22199    |    .    |   *.    |
1.00000   0.75243    0.24757    |    .    |   *.    |
1.00000   0.67971    0.32029    |    .    |    *.   |
1.00000   0.67693    0.32307    |    .    |    *.   |
1.00000   0.69179    0.30821    |    .    |    *.   |
0.00000   0.76561   -0.76561    |*    .    |    .    |
1.00000   0.67918    0.32082    |    .    |    *.   |
1.00000   0.73916    0.26084    |    .    |   *.    |
1.00000   0.89765    0.10235    |    .    |*   .    |
1.00000   0.80180    0.19820    |    .    |*   .    |
1.00000   0.79023    0.20977    |    .    |*   .    |
1.00000   0.82349    0.17651    |    .    |*   .    |
1.00000   0.84845    0.15155    |    .    |*   .    |
1.00000   0.82972    0.17028    |    .    |*   .    |
1.00000   0.84894    0.15106    |    .    |*   .    |
0.00000   0.83381   -0.83381    |*    .    |    .    |
1.00000   0.90241    0.09759    |    .    |*   .    |
1.00000   0.88119    0.11881    |    .    |*   .    |
1.00000   0.82376    0.17624    |    .    |*   .    |
1.00000   0.80414    0.19586    |    .    |*   .    |
1.00000   0.80425    0.19575    |    .    |*   .    |
1.00000   0.82562    0.17438    |    .    |*   .    |
1.00000   0.84521    0.15479    |    .    |*   .    |
1.00000   0.79309    0.20691    |    .    |*   .    |
0.00000   0.70667   -0.70667    |*    .    |    .    |
1.00000   0.89384    0.10616    |    .    |*   .    |
1.00000   0.85911    0.14089    |    .    |*   .    |
1.00000   0.80916    0.19084    |    .    |*   .    |
1.00000   0.92547    0.07453    |    .    |*   .    |
1.00000   0.75827    0.24173    |    .    |   *.    |
1.00000   0.75293    0.24707    |    .    |   *.    |
1.00000   0.79911    0.20089    |    .    |*   .    |
1.00000   0.77590    0.22410    |    .    |   *.    |
1.00000   0.77570    0.22430    |    .    |   *.    |
1.00000   0.79382    0.20618    |    .    |*   .    |
1.00000   0.79669    0.20331    |    .    |*   .    |
1.00000   0.79669    0.20331    |    .    |*   .    |
1.00000   0.87543    0.12457    |    .    |*   .    |
0.00000   0.81415   -0.81415    |*    .    |    .    |
1.00000   0.66876    0.33124    |    .    |    *.   |
1.00000   0.79874    0.20126    |    .    |*   .    |
1.00000   0.85496    0.14504    |    .    |*   .    |
```

```
1.00000   0.81024    0.18976       |    . | *  .   |
1.00000   0.83919    0.16081       |    . | *  .   |
1.00000   0.90064    0.09936       |    . |*   .   |
1.00000   0.89158    0.10842       |    . |*   .   |
1.00000   0.80731    0.19269       |    . | *  .   |
1.00000   0.87485    0.12515       |    . | *  .   |
1.00000   0.77902    0.22098       |    . | *  .   |
1.00000   0.84275    0.15725       |    . | *  .   |
1.00000   0.87964    0.12036       |    . | *  .   |
0.00000   0.84713   -0.84713       | *    .  |     |
1.00000   0.81132    0.18868       |    . | *  .   |
1.00000   0.79337    0.20663       |    . | *  .   |
1.00000   0.79928    0.20072       |    . | *  .   |
1.00000   0.86172    0.13828       |    . | *  .   |
1.00000   0.84569    0.15431       |    . | *  .   |
1.00000   0.70621    0.29379       |    . | *.    |
1.00000   0.80622    0.19378       |    . | *  .   |
1.00000   0.79475    0.20525       |    . | *  .   |
1.00000   0.89579    0.10421       |    . |*   .   |
1.00000   0.70089    0.29911       |    . | *.    |
0.00000   0.74372   -0.74372       | *    .  |     |
1.00000   0.75052    0.24948       |    . | *  .   |
1.00000   0.70792    0.29208       |    . | *.    |
1.00000   0.78615    0.21385       |    . | *  .   |
1.00000   0.75577    0.24423       |    . | *.    |
1.00000   0.65302    0.34698       |    . | *.    |
1.00000   0.74396    0.25604       |    . | *.    |
1.00000   0.66568    0.33432       |    . | *.    |
1.00000   0.82952    0.17048       |    . | *  .   |
1.00000   0.78180    0.21820       |    . | *  .   |
1.00000   0.78486    0.21514       |    . | *  .   |
1.00000   0.85320    0.14680       |    . | *  .   |
1.00000   0.89656    0.10344       |    . |*   .   |
1.00000   0.78813    0.21187       |    . | *  .   |
1.00000   0.86843    0.13157       |    . | *  .   |
1.00000   0.75212    0.24788       |    . | *  .   |
1.00000   1.00000    0.00000       |    . | *  .   |
0.00000   0.77820   -0.77820       | *    .  |     |
1.00000   0.77820    0.22180       |    . | *  .   |
1.00000   0.82143    0.17857       |    . | *  .   |
1.00000   0.76317    0.23683       |    . | *  .   |
1.00000   0.88444    0.11556       |    . |*   .   |
1.00000   0.84874    0.15126       |    . | *  .   |
1.00000   0.76243    0.23757       |    . | *  .   |
1.00000   0.82852    0.17148       |    . | *  .   |
1.00000   0.70104    0.29896       |    . | *.    |
1.00000   0.78126    0.21874       |    . | *  .   |
1.00000   0.75846    0.24154       |    . | *  .   |
1.00000   0.60678    0.39322       |    . | *.    |
1.00000   0.56327    0.43673       |    . | .*    |
1.00000   0.88268    0.11732       |    . | *  .   |
1.00000   0.86624    0.13376       |    . | *  .   |
1.00000   0.83480    0.16520       |    . | *  .   |
1.00000   0.88454    0.11546       |    . |*   .   |
1.00000   0.88453    0.11547       |    . |*   .   |
1.00000   0.95059    0.04941       |    . |*   .   |
1.00000   0.90355    0.09645       |    . |*   .   |
1.00000   0.94092    0.05908       |    . |*   .   |
1.00000   0.84978    0.15022       |    . | *  .   |
1.00000   0.84248    0.15752       |    . | *  .   |
```

```
1.00000    0.82793    0.17207      |    .   |* .    |
1.00000    0.86303    0.13697      |    .   |* .    |
1.00000    0.88207    0.11793      |    .   |* .    |
1.00000    0.89805    0.10195      |    .   |* .    |
1.00000    0.89107    0.10893      |    .   |* .    |
1.00000    0.76278    0.23722      |    .   | * .   |
1.00000    0.84663    0.15337      |    .   | * .   |
1.00000    0.89433    0.10567      |    .   |* .    |
1.00000    0.72167    0.27833      |    .   |  *.   |
1.00000    0.84666    0.15334      |    .   | * .   |
1.00000    0.79509    0.20491      |    .   | * .   |
1.00000    0.83508    0.16492      |    .   | * .   |
1.00000    0.92232    0.07768      |    .   |* .    |
1.00000    0.83058    0.16942      |    .   | * .   |
```

APPENDIX V

SCHOOL OF POST GRADUATE STUDIES
BAYERO UNIVERSITY, KANO
P.M.B 1002, KANO NIGERIA

I am a post graduate student of M.sc Economics from the above mentioned university. I am conducting a research on the "Determinants of Food Insecurity in Northern Nigeria (A case of Jigawa State)".

Respondents are assured that any information given will be strictly used for academic purpose only.

Please specify or tick [   ] appropriately as the case may be.

SECTION A
GENERAL INFORMATION

1. What is the L.G.A. of the Household? _____

2. What is the sex of the household?
   a). Male [   ]   b) Female [   ]

3. What is the Age of the Household?
   a) <29 [   ]   b) 30-49 [   ]   c)50-69 [   ]   d) 70 & above [   ]

4. What is your Marital Status?
   a). Single [   ]   b) Married [   ]

5. If married, how many wives do you have?
   a) One Wife [   ]   b) Two wives [   ]
   c) Three Wives [   ]   d)Four wives [   ]

6. What is the size of your Household (no. of dependants)?
   a) 1-5 [   ]   b) 6-10 [   ]   c) 11-15 [   ]
   d)16-20 [   ]   e) 20 + [   ]   f) No Children [   ]

7. What is your educational Qualification?
   a) Informal Education [   ] b)Primary Education [   ]
   c) Secondary Education [   ] d)OND/NCE [   ]
   e) BSC/HND [   ]   f) Others, specify_____

8. What is your Occupation?
   a) Farming [   ] b)Trading/Business [   ]
   c) Salary/Wage Earner [   ]   d) Others, specify_____

9. Do you own a Farm land?
   a) Yes [____] b) No [____]

10. If yes, what is the estimate of your Farm Size?
    a) 1.5 acres [____] b) 2 acres [____] c) 2.5 acres [____]
    b) 3 acres [____] e) 3.5 acres [____] f) Others, specify_____

11. What are the sources of your farm inputs? a) Market [____]
    b) Government [____] c) Individual [____] d) Others _____

12. Do you receive any support from Government?
    a) Yes [____] b) No [____]

13. If yes, what kinds of support do you received?
    a) Fertilizer [____] b) Seedling [____] c) Others _____

14. What is the estimate of your Monthly Income?
    a) < ₦ 10, 000 [____] b) ₦ 10, 000 – ₦ 39, 000 [____]
    c) ₦ 40, 000 – ₦ 69, 000 [____] d) ₦ 70, 000 – ₦ 99, 000 [____]
    e) ₦ 100, 000 + [____] f) No Income [____]

15. Do you have a secondary occupation apart from your occupation?
    a) Yes [____] b) No [____]

16. If yes, specify_____

SECTION B
FOOD CONSUMED BY THE HOUSEHOLDS WITHIN 24 HOURS

1. Do you normally have food for breakfast?
   a) Yes [____] b) No [____]

2. If yes, specify the kind of food taken for breakfast_____

3. Are you satisfied with the food you take for breakfast?
   a) Yes [____] b) No [____]

4. If yes, what are your reasons? _____

5. What is the estimated cost of your Breakfast?
   a) < ₦157 [____] b) ₦157 – ₦ 300 [____]
   b) ₦301 – ₦ 470 [____] d) ₦ 471 and Above_____

6. Do you normally have food for lunch?
   b) Yes [____] b) No [____]

7. If yes, specify the kind of food taken for lunch?_____

8. Are you satisfied with the food you take for lunch?   a)
   Yes   [_____] b) No   [_____]

9. If yes, what are your reasons? _____

10. What is the estimated cost of your Lunch?
    a) < ₦ 157   [_____] b)₦157 – ₦ 300   [_____]
    b) ₦301 – ₦ 470   [_____] d)₦ 471 & above _____

11. Do you normally have food for Dinner?
    a)   Yes [_____]   b) No [_____]

12. If yes, specify the kind of food taken for Dinner? _____

13. Are you satisfied with the food you take for Dinner?
    a)   Yes [_____]   b) No [_____]

14. If yes, what are your reasons? _____

15. What is the estimated cost of your Dinner?
    a) < ₦ 157   [_____] b) ₦157 – ₦ 300 [_____]
    c) ₦301 – ₦ 470 [_____] d) > ₦ 471 & above _____

16. What do you think causes Food Insecurity?
    a) Family size [_____] b) Illiteracy [_____] c) Poverty [_____]
    b) Others _____

17. Does household income improve Food Security?
    a) Yes [_____]   b) No [_____]

Thank You!

Printed in Great Britain
by Amazon

68212525R00068